ELECTRICAL TECHNOLOGY

ELECTRICAL TECHNOLOGY

Amit Charan

ANMOL PUBLICATIONS PVT. LTD.
NEW DELHI-110 002 (INDIA)

ANMOL PUBLICATIONS PVT. LTD.
Regd. Office: 4360/4, Ansari Road, Daryaganj,
New Delhi-110002 (India)
Tel.: 23278000, 23261597, 23286875, 23255577
Fax: 91-11-23280289
Email: anmolpub@gmail.com
Visit us at: www.anmolpublications.com

Branch Office: No. 1015, Ist Main Road, BSK IIIrd Stage
IIIrd Phase, IIIrd Block, Bangalore-560 085 (India)
Tel.: 080-41723429 • Fax: 080-26723604
Email: anmolpublicationsbangalore@gmail.com

Electrical Technology

First Edition, 2010

ISBN 978-81-261-4600-0

PRINTED IN INDIA

Printed at Mehra Offset Press, Delhi.

Contents

Preface

The textbook of Electrical Technology is a unique and modern textbook for the students of Diploma and Degree level of UPTu and other State Universities. The book presents the material in an integrated manner and logical sequence thereby covering all integrated course material as per the syllabus. The book comprises of solved examples, exercises and detail explanation of the topics as required. Apart from learning d.c and a.c circuit analysis both under steady state and transient conditions, students will learn basic working principles and analysis of transformer, d.c motors and induction motor. The working principles of some popular and useful indicating measuring instruments are presented.

The objective of this book is to give the detailed explanations about the various Circuits and their Application which gives the brief knowledge to the students in the way they required. All Objective Tests have been recast and extended where required. The main aim of this textbook is to provide a comprehensive coverage in a clear and well illustrated format.

Author

Chapter 1

D.C. Circuit Analysis

SOURCE

The source of electric power in our houses, offices, class rooms etc is the AC source. A single source works as voltage and current source. Systems like Radio, TV, Music etc have inbuilt dc powers supply that converts ac in to dc to supply power to these and we enjoy programme of our choices.

The value of voltage (v) or current (i) of ac source varies sinusoidally with time t and can be written as

$$v(t) = V_{max} \sin \omega t \quad or \quad v(t) = V_{max} \cos \omega t$$

$$i(t) = I_{max} \sin \omega t \quad or \quad i(t) = I_{max} \cos \omega t$$

V_{max} and I_{max} the peak values of voltage and current

and $\omega = 2\pi f$ is the anugar frequency of the sourse.

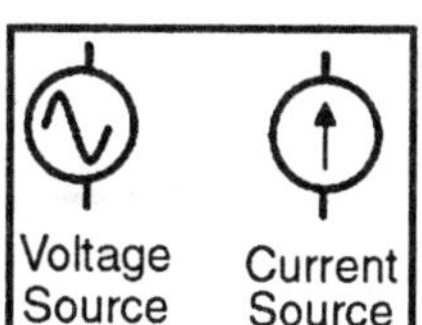

For simplicity we write $v(t)$ as v and $i(t)$ as i.

The variation of output of voltage/current source is represented below

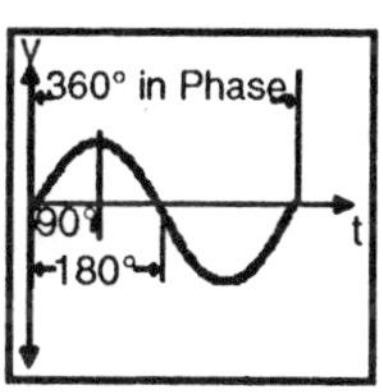

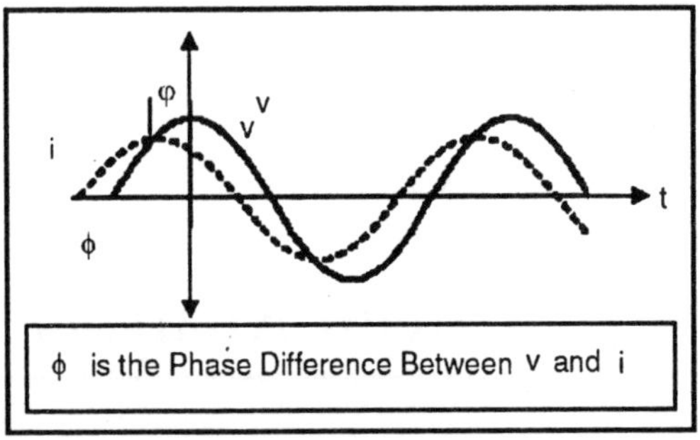

AC SOURCE AND RESISTOR(R)

We shall study the phase difference between the current in the resistor and the voltage across the R.

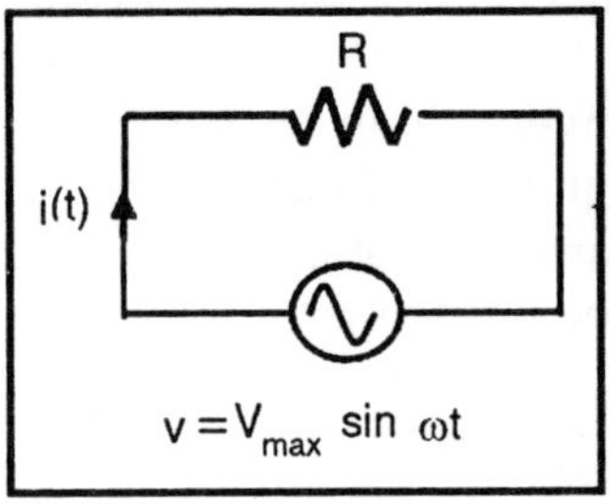

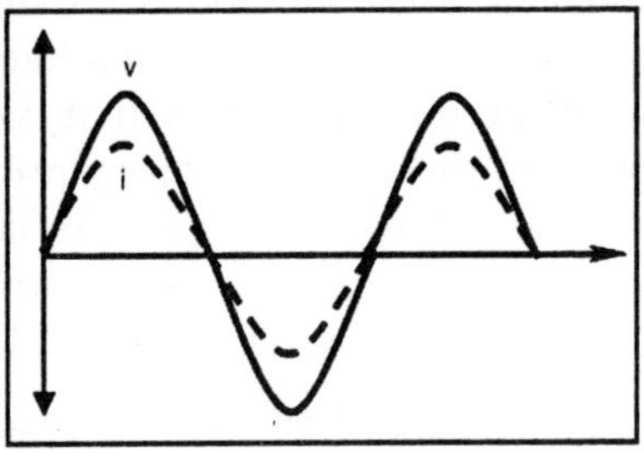

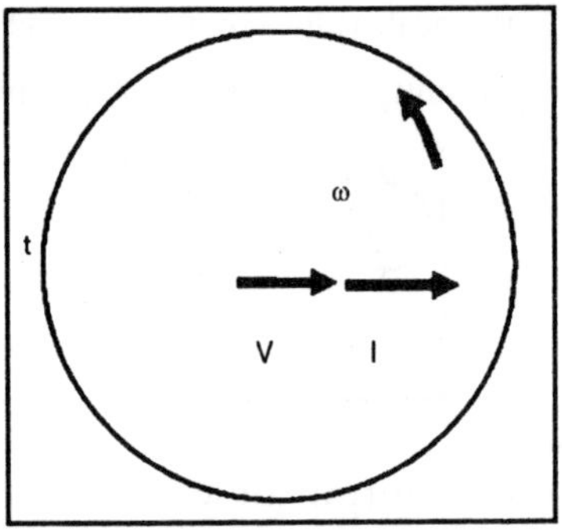

$$v = V_{max} \sin \omega t$$

Let instantaneous value of potential from source is given by,

$$v = V_{max} \sin \omega t$$

Instantaneous value of current from source can be obtained as,

$$i = \frac{v}{R} = \frac{V_{max} \sin \omega t}{R}$$
$$= I_{max} \sin \omega t$$

Where $\frac{V_{max}}{R} = I_{max}$ (max. value of current)

The ratio,

$$V_{max}/I_{max}$$

is called the *reactance* due to resistor,

$$X_R.$$

$$X_R = R \text{ is called resistance}$$

Above Eqs; indicate that v and i in the resistor are in the phase.

The average values of v and i in one cycle are zero. Therefore, average value of i and v are given by ttheir *rms* values. The effective or *rms* value of voltage in one cycle is given by,

$$V_{rms} = \sqrt{\frac{\int_0^T v^2 dt}{T}} = \frac{V_{\max}}{\sqrt{2}} = 0.707\, V_{\max}$$

Similarly *rms* value of current can be obtained as,

$$I_{rms} = \frac{V_{max}}{\sqrt{2}}$$

AC Source and inductance (L):

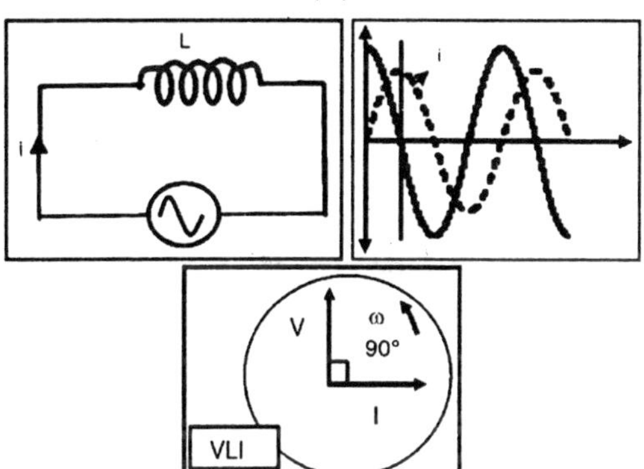

$$v = V_{max} \sin \omega t$$

Assume the resistance of the coil is negligible.

Let,

$$i = I_{max} \sin wt,$$

the changing current sets up a back- *emf* in the coil, given by,

$$\varepsilon = -L \frac{di}{dt}$$

To maintain the current, the applied voltage must be equal to the back-*emf i.e.,*

$$v = L\frac{di}{dt}$$

substituting the value of $i(t)$ in above Eq;

$$X_C = \frac{V_{\max}}{I_{\max}} = \frac{1}{\omega C}$$

we get,

$$v = LI_{\max}\omega \cos \omega t \quad or$$

where,

$$V_{max} = \omega L I_{max}$$

$$\frac{V_{max}}{I_{max}} = \omega L$$

is called reactance due to inductor and is written as X_L.

Reactance of,

$$L, X_L = w L$$

The voltage across the L leads the current through the L by 90° or the current through the L lags the voltage across the L by 90°.

Problem

An inductor has a 54.0-Ω reactance at 60.0 Hz. What is the maximum current if this inductor is connected to a 50.0-Hz source that produces a 100-V rms voltage?

Sol

At 50.0 Hz, S,

$$X_L = 2\pi(50.0\text{ Hz})L = 2\pi(50.0\text{ Hz})\left(\frac{X_L|_{60.0\text{ Hz}}}{2\pi(60.0\text{ Hz})}\right) = \frac{50.0}{60.0}(54.0\ \Omega)$$

$= 45.0\ \Omega$

$$I_{\max} = \frac{\Delta V_{\max}}{X_L} = \frac{\sqrt{2}(\Delta V_{\text{rms}})}{X_L} = \frac{\sqrt{2}(100\text{ V})}{45.0\ \Omega} = \boxed{3.14\text{ A}}$$

AC SOURCE AND CAPACITOR C

The charge (+, -) will be flowing from terminals of the supply to the plates of capacitor in the circuit and deposit on plates that results potential difference across the plates of capacitor. The dc current does not pass through a capacitor since it offers infinite reactance to dc current.

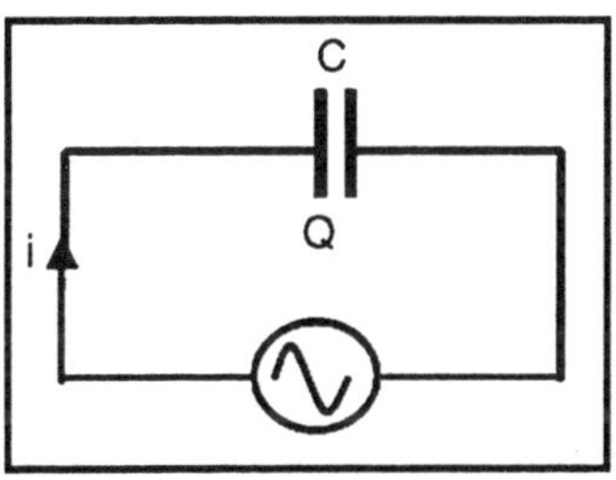

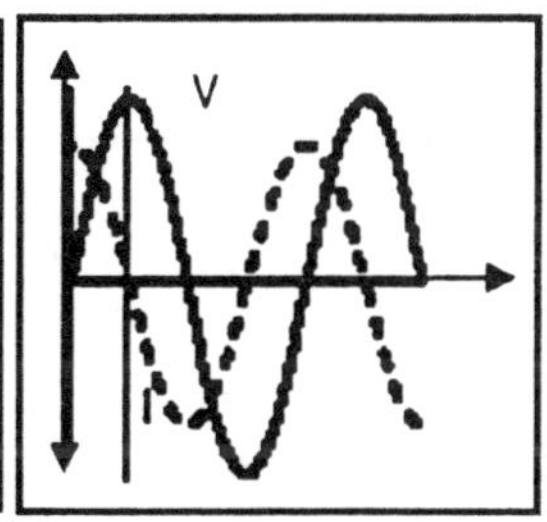

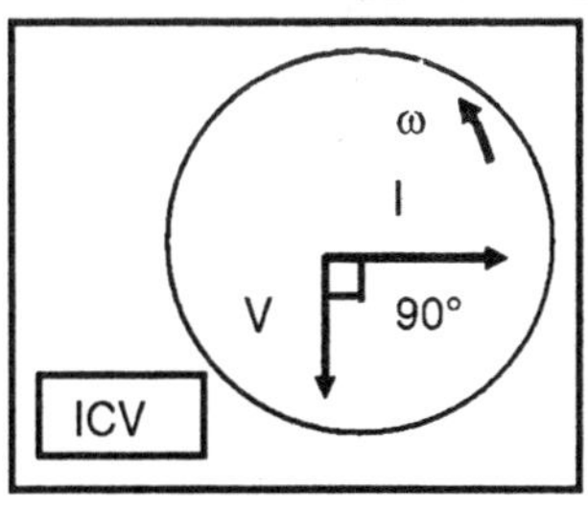

$$v = V_{\max}\sin\omega t$$

Given,

$$v = V_{\max}\sin wt$$

The Charge Q in the capacitor with time t is given by,

$$Q = C\,v = C\,V_{max}\sin wt$$

The current in the circuit is,

$$i = \frac{dQ}{dt} = \frac{d(CV_{\max}\sin\omega t)}{dt} = \omega CV_{\max}\cos\omega t$$

$$i(t) = \omega C V_{max} \sin\left(\omega t + \frac{\pi}{2}\right)$$

From above Eq; we can infer that current leads the voltage by 90°. Or the voltage across the C lags the current through C by 90°.

The $i(t)$ can be written as,

$$i(t) = I_{max} \cos \omega t$$

From above Eq;

$$I_{max} = \grave{U}\, CV_{max}$$

Hence reactance of capacitor,

$$X_C = \frac{V_{max}}{I_{max}} = \frac{1}{\omega C}$$

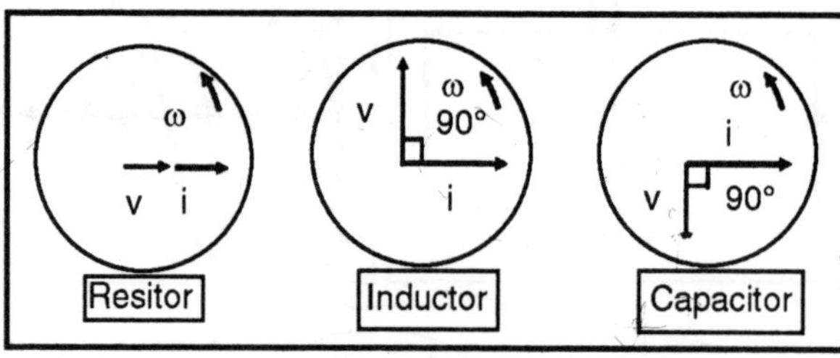

Fig. Simple Diagram to Understand the Phase Difference Between v and i

Problem

What maximum current is delivered by an AC source with V_{max} = 48.0 V and f = 90.0 Hz when connected across a 3.70-μF capacitor?

Sol:

$$I_{max} = (\Delta V_{max})\omega C = (48.0\text{ V})(2\pi)(90.0\text{ s}^{-1})(3.70\times10^{-6}\text{ F})$$

$$= \boxed{100\text{ mA}}$$

RLC SERIES CIRCUIT

There are many ways of analysing AC circuits but we will

use a simple method using a phasor diagram. This is like a vector diagram and voltages and currents are represented by lines with arrows drawn on them.

We have to find the impedance of the circuit, current in the circuit, resonating frequency and phase difference between *v* and *i* in the ckt.

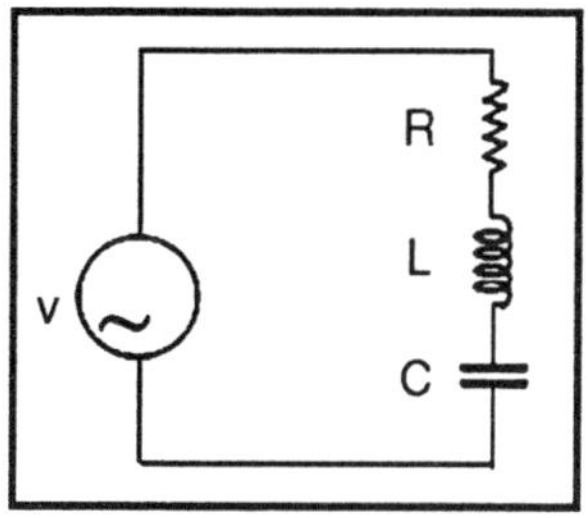

Fig. Phasor

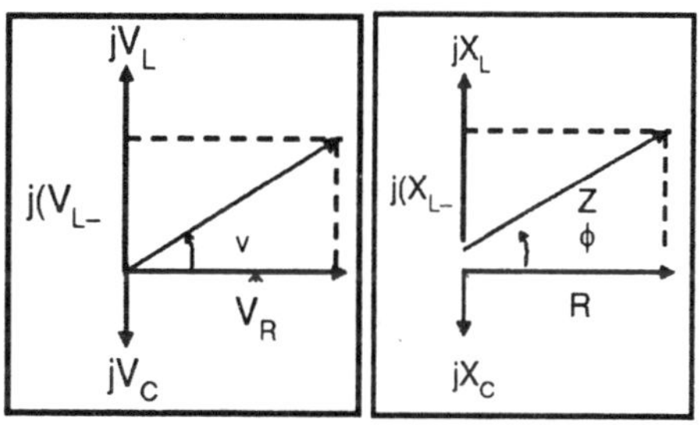

- *The impedance Z:* Z of the circuit $Z = R + j\,(X_L - X_C)$,

$$Z = \sqrt{R^2 + (X_L - X_C)^2}$$

- *Resultant voltage:*

$$V = \sqrt{V_R^2 + (V_L - V_C)^2}$$

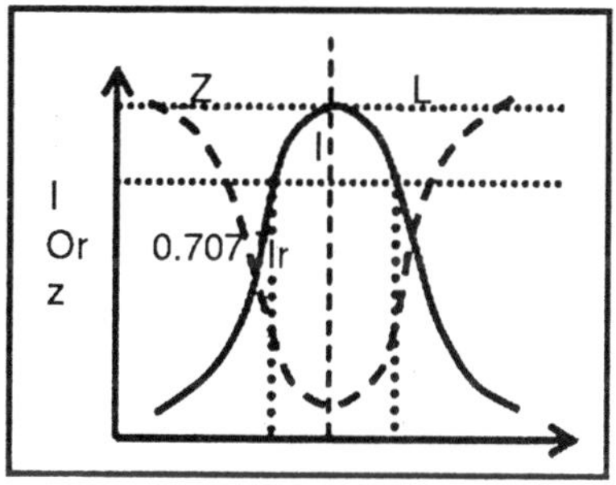

- *Current I:* The current,

$$I = \frac{V}{Z} = \frac{V}{\sqrt{R^2 + (X_L - X_C)^2}}$$

- *Phase differencef:* The phase difference between v and i.

$$\phi = \tan^{-1}\left(\frac{X_L - X_C}{R}\right) = \tan^{-1}\left(\frac{V_L - V_C}{V_R}\right)$$

- *Resonating Frequency fx:* At resonance, $X_L = X_C$, there fore resonating frequency is given by

$$f_r = \frac{1}{2\pi\sqrt{LC}}$$

and

$$Z = R$$

At resonance the impedance and resultant voltage are minimum and current is maximum.

POWER IN AN AC CIRCUIT

The average power dissipated in any element can also be written as,

$$P_{av} = \int_0^T \frac{v \times i}{T} dt$$

Where v is the voltage across the element and i is the current flowing through it. For the inductor and the capacitor the average power dissipated is given by,

$$P_{av} = \int_0^T \frac{V_{\max} I_{\max} \sin \omega t \cos \omega t}{T} dt$$

The value of integral,

$$\int_0^T \frac{\sin \omega t \ \cos \omega t dt}{T} dt = 0$$

Therefore, power dissipated (loss) in pure Inductor L and capacitor C are Zero. This is the power supplied by the source. If there is a resistor in the circuit, the power is lost as heat. In L and C the average power supplied by the source is zero. In RLC circuit there is a phase difference f between $v(t)$ and $i(t)$.

Let $v = V_{max} \sin wt$
Then $i = I_{max} \sin (wt \pm f)$
$= I_{max} (\sin wt \cos f \pm \cos wt \sin f)$
Power dissipated in RLC circuit:

$$P_{av} = \frac{1}{T}\int_0^T V_{max} I_{max} \left(\sin^2 \omega t \cos\phi \, dt \pm \sin\omega t \cos\omega t \sin\phi dt\right)$$

The integral of the first factor (time average) is ½ cos ϕ and the integral of the second factor (time average) is 0. Therefore, average power P_{av} delivered by the source,

$$P_{av} = \frac{V_{max}\, I_{max}}{2} \cos\phi$$

$= V_{rms} I_{rms} \cos f = I_{rms}{}^2 R$

Problem

A resonating circuit shown in Fig. has, C=300 pF, L= 220 μH R= 20Ω. If supply voltage is v(t)=10 sin*wt* V. Calculate,

- Resonating frequency
- Impedance of ckt Z_s at ω=300 k rad/sec
- Impedance at resonance,
- Voltage across each component at resonance
- Q factor and bandwidth of ckt.

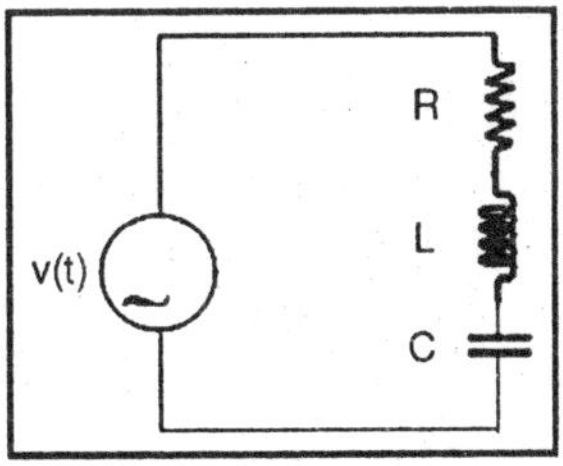

Sol

$$f_r = \frac{1}{2\pi\sqrt{LC}}$$

$$= \frac{1}{2\pi\sqrt{220\times 10^{-6}\; 300\times 10^{-12}}} = 620kHz$$

– impedance at w=300 k rad/sec

$$Z_s = \left[R^2 + \left(X_C - X_L\right)^2\right]^{1/2}$$

$$= \sqrt{20^2 + (0.6 - 6.6)10^6} = 2.5k\Omega$$

- Impedance at resonance Z = R = 20W
- Currsent at resonance

$$I_r = \frac{V}{R} = \frac{10}{20} = 0.5A$$

- Potentail across R, L, and C

$$V_R = IR = 10V,$$

$$V_L = I\omega_r L = 428V$$

$$V_C = I\frac{1}{\omega_r C} = 428V$$

- Quality factor Q

$$Q = \frac{X_L}{R} = \frac{\omega_r L}{R} = \frac{2\times\pi\times620\times10^3\times220\times10^{-6}}{20} = 42.8$$

Problem

A series RLC circuit has a resistance of 45 Ω and an impedance of 75 Ω. What average power is delivered to the circuit when V_{rms} = 210 V.

Sol

$$\mathrm{P} = (V_{rms})I_{rms}\cos\phi$$

$$(X_L - X_C) = \sqrt{(75.0\ \Omega)^2 - (45.0\ \Omega)^2} = 60.0\ \Omega$$

$$\phi = \tan^{-1}\left(\frac{X_L - X_C}{R}\right) = \tan^{-1}\left(\frac{60.0\ \Omega}{45.0\ \Omega}\right) = 53.1^\circ$$

$$I_{rms} = \frac{V_{rms}}{Z} = \frac{210\ \mathrm{V}}{75.0\ \Omega} = 2.80\ \mathrm{A}$$

$$P = (V_{rms}) I_{rms} \cos\phi = (210\text{ V})(2.80\text{ A})\cos(53.1¢^{a})$$

353 W

A network of passive elements and sources is a circuit.

Analysis: To determine currents or voltages in various elements (effects) due to various sources (cause).

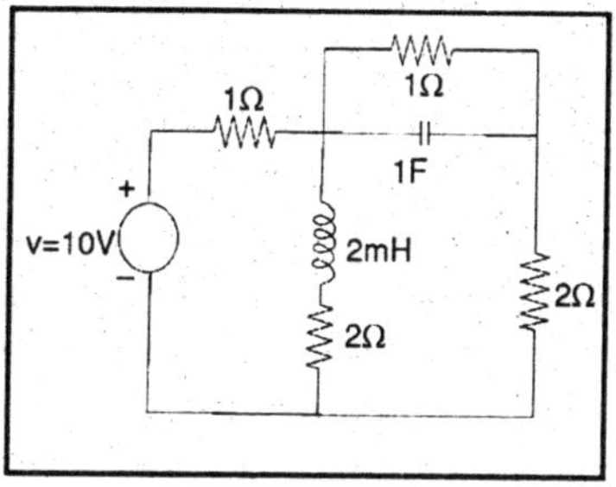

In circuit 4.1 v=10V all the time.

Expected that all current (voltage) in (across) the elements is constant.

INDUCTOR

$$v = L\frac{di}{dt}$$

$$\frac{di}{dt} = 0$$

Implies i = constant.

Hence $v = 0$

Inductors act as a short cicuit for DC inputs. This would not be the case if I put a switch across a source.

Capacitor:

$$i = C\frac{dv}{dt} \text{ as } \frac{dv}{dt} = 0 \text{ (expected), i=0}$$

Thus capacitor acts as open circuit for DC analysis.

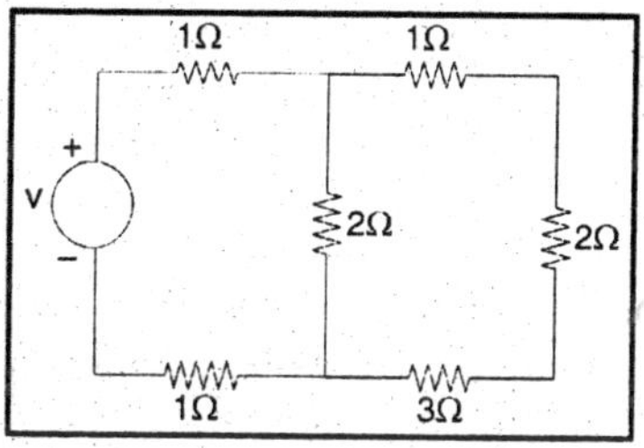

The resultant circuit will be as shown in Fig..

Analysis

To find currents in all branches, voltage across all branches.

We can use Kirchoff's law (voltage and current). For as many independent equations as number of unknown variables. Solve the simultaneous equations, and get the result.

V_{ab} = *Voltage drop from'a'to'b'.Therefore,* $V_{ab} = -V_{ba}$

I_{ab} =*Current in branch ab in the direction from 'a' to 'b'. Here* $I_{ab} = -I_{ba}$

Hence

$$V_{fa} = -v$$
$$V_{ab} = I_{ab} \times 1 = I_{ab}$$
$$V_{be} = I_{be} 2$$
$$V_{ef} = I_{ef} 1$$
$$V_{bc} = I_{bc} 1$$
$$V_{cd} = I_{cd} 2$$
$$V_{de} = I_{de} 3$$
$$V_{ab} + V_{be} + V_{ef} = v$$
$$I_{fa} = I_{ab}$$
$$I_{ef} = I_{fa}$$
$$I_{bc} = I_{cd}$$
$$I_{cd} = I_{de}$$
$$V_{ab} + V_{bc} + V_{cd} + V_{de} + V_{ef} = v$$
$$I_{bc} + V_{be} = I_{ab}$$

Note: In a circuit with N nodes, the number of branches B will alwasy be$^{(N - 1) + L}$, where L are maximum number of independent closed paths possible in the circuit.

Hence, we can always form B equations using Ohm's law1, $^{(N - 1)}$ equations using KCL, L equations using KVL. Hence in total 2B equations can be formed, which are sufficient to solve for 2B variables (voltage and current in each branch).

SIMPLIFYING

Loop currents method

We do away with branch currents and define loop

currents. The branch currents can be written in terms of loop currents once all the loop currents passing through the branch and their directions are known.

The branch voltages can always be written using Ohm's law and branch current written in terms of loop currents. So now our objective is to find loop currents.

For this we choose maximum number of independent loops Figure and apply KVL in them.

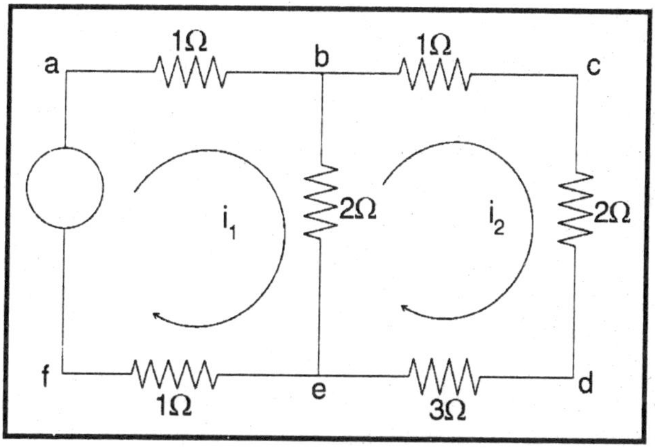

If i1 and i2 are known, voltages across all elements can befound.

Make two independent equations:

For loop abef:

$$v - i_1 \times 1 - (i_1 - i_2) \times 2 - i_1 \times 1 = 0$$

$$v = i_1(1 + 2 + 1) - 2i_2$$

For loop bcde:

$$-(i_2 - i_1)2 - 1i_2 - 2i_2 - 3i_3 = 0$$

$$0 = -2i_2 + i_2(2 + 1 + 2 + 3)$$

Use any technique to solve these (such as using matrices). *We get:*

$$i_1 = \frac{2}{7} \, Amperes$$

$$i_2 = \frac{2}{7} \, Amperes$$

NODAL VOLTAGE METHOD

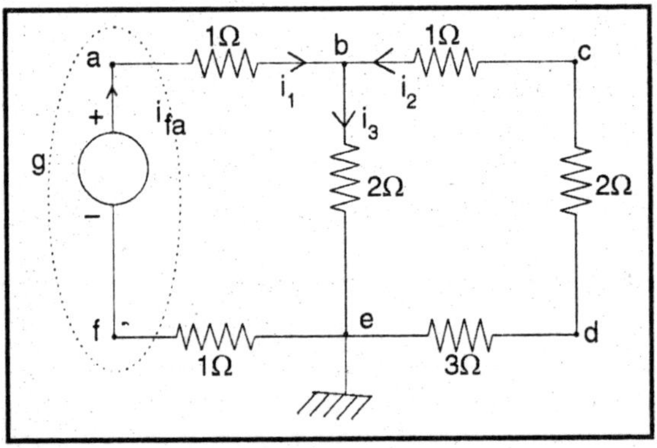

INDEPENDENT NODES

One of the nodes in circuit need to be considered as reference node. Hence its node potential is zero. For other nodes, nodal voltage is potential differetial w.r.t. to reference node.

The nodes are called independent nodes. In general for N node network, $N - 1$ nodes will be independent.

At node b:

$$i_1 + i_2 - i_3 = 0\,.$$

Similarly, other equations are:

$$\frac{E_a - E_b}{1} - \frac{E_c - E_b}{1} - \frac{E_b}{2} = 0$$

$$\frac{E_a - E_b}{1} = i_{fa}$$

$$E_a = E_f + v$$

$$\frac{-E_f}{1} = i_{fa}$$

$$\frac{E_c - E_b}{1} = \frac{E_d - E_c}{2}$$

$$\frac{-E_d}{3} = \frac{E_d - E_c}{2}$$

These are six equations, in six unknowns. Thus can be solved for a unique solution. One can make a supernode and use KCL combining the nodes nodes a and f. We also make extra equations for potential difference between two nodes.

With supernode, no. of equations is equal to no. of independent nodes whose voltage w.r.t. reference needs to be determined.

CURRENT SOURCES IN LOOP ANALYSIS

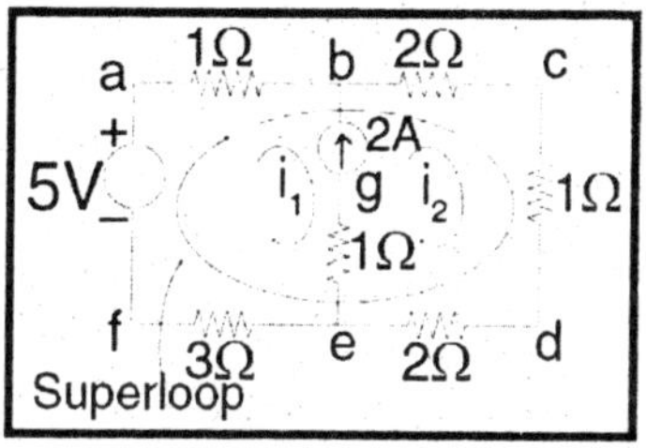

Fig. Current sources in Loop current analysis

Using KVL for loop 1 in Figure

$$5 - i_1 - v_{bg} - (i_1 - i_2)1 - i_1 3 = 0$$

The other equations are:

$$i_1 - i_2 = 2$$

$$v_{bg} - i_2 2 - i_2 1 - i_2 2 - (i_2 - i_1) = 0$$

The first and the third equation can be combined for taking care of V_{bg}. This can be done by making superloop for writing KVL.

GRAPH

For analysing circuits efficiently.

Loop current method

One can form a spanning tree from graph such that current sources are in links (Those elements which do not form part of the tree). Each link when added to the tree gives a loop. All voltage sources should be kept in branches of tree.

For example, refer to the following figures.

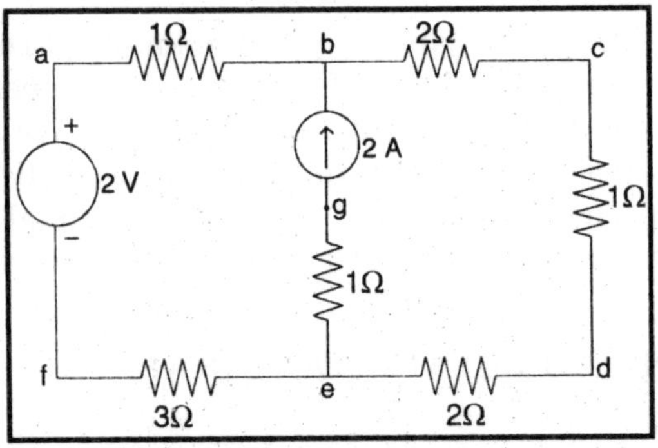

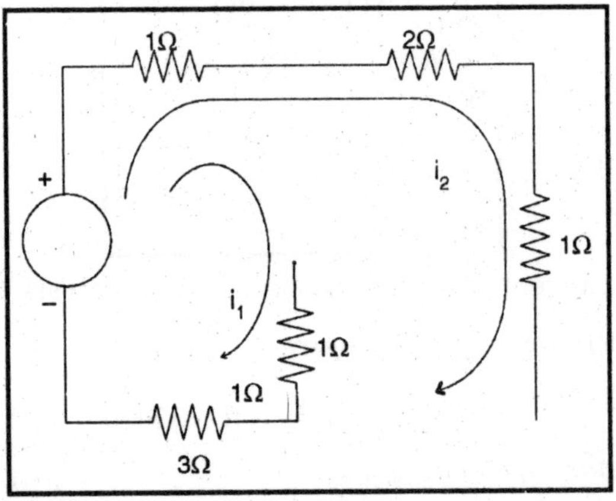

Node Voltages

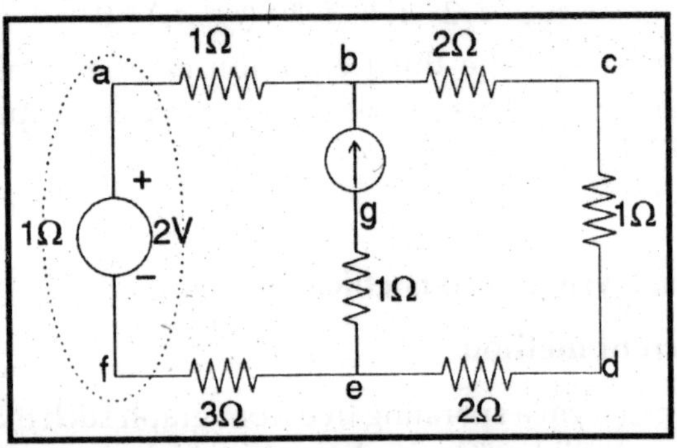

In the above figure $V_{af} = 2V$. There are five unknown node voltages in the above circuit, namely, V_{ab}, V_{fe}, V_{bc}, V_{cd} and V_{ge}.

Correspondingly, we have five equations. Note that we can merge af into one supernode

$$V_{ab} + V_{fe} / 3 = 0$$

$$V_{ab} + V_{bc} / 2 + 2 = 0$$

$$V_{bc} / 2 - V_{cd} = 0$$

$$V_{cd} - V_{de} / 2 = 0$$

$$V_{ge} + 2 = 0$$

The second equation follows from looking at node b, while the third one from doing the same at node c.

NETWORK THEOREMS

Anyone who's studied geometry should be familiar with the concept of a theorem: a relatively simple rule used to solve a problem, derived from a more intensive analysis using fundamental rules of mathematics.

At least hypothetically, any problem in math can be solved just by using the simple rules of arithmetic (in fact, this is how modern digital computers carry out the most complex mathematical calculations: by repeating many cycles of additions and subtractions!), but human beings aren't as consistent or as fast as a digital computer. We need "shortcut" methods in order to avoid procedural errors.

In electric network analysis, the fundamental rules are Ohm's Law and Kirchhoff's Laws. While these humble laws may be applied to analyse just about any circuit configuration (even if we have to resort to complex algebra to handle multiple unknowns), there are some "shortcut" methods of analysis to make the math easier for the average human.

As with any theorem of geometry or algebra, these network theorems are derived from fundamental rules. In this chapter, I'm not going to delve into the formal proofs of any of these theorems.

If you doubt their validity, you can always empirically test them by setting up example circuits and calculating values using the "old" (simultaneous equation) methods versus the "new" theorems, to see if the answers coincide.

OBJECTIVES

The objective of the DC Network Theorems Circuit Board and Course is to familiarize students with the Kirchhoff's Current and Voltage Laws, Kirchhoff's Laws Combined, Kirchhoff's Solution with Two Sources, Superposition, Thevenin Circuits, Thevenizing a Bridge Circuit, Thevenin/ Norton Conversion, and Delta to Y or Y to Delta. The course provides comprehensive, hands-on instruction in the terminology, principles, and applications of the circuitry used in DC networks.

NORTON'S THEOREM

Norton's Theorem states that it is possible to simplify any linear circuit, no matter how complex, to an equivalent circuit with just a single current source and parallel resistance connected to a load. Just as with Thevenin's Theorem, the qualification of "linear" is identical to that found in the Superposition Theorem: all underlying equations must be linear (no exponents or roots).

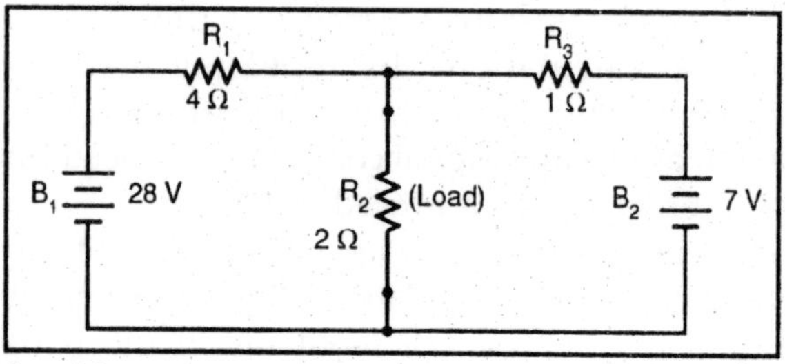

...after Norton conversion...

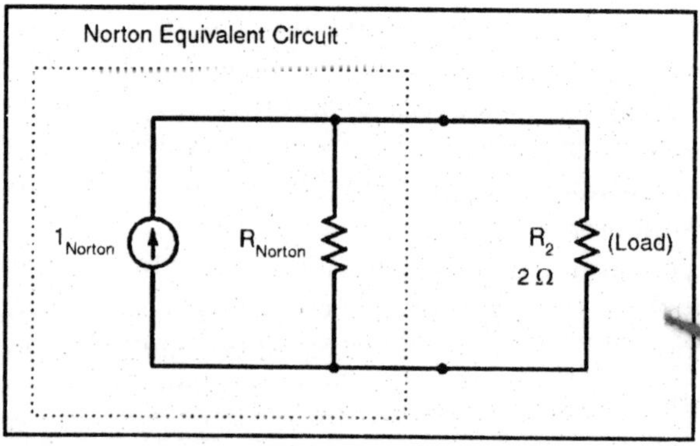

Remember that a *current source* is a component whose job is to provide a constant amount of current, outputting as much or as little voltage necessary to maintain that constant current.

As with Thevenin's Theorem, everything in the original circuit except the load resistance has been reduced to an equivalent circuit that is simpler to analyse. Also similar to Thevenin's Theorem are the steps used in Norton's Theorem to calculate the Norton source current (I_{Norton}) and Norton resistance (R_{Norton}).

As before, the first step is to identify the load resistance and remove it from the original circuit:

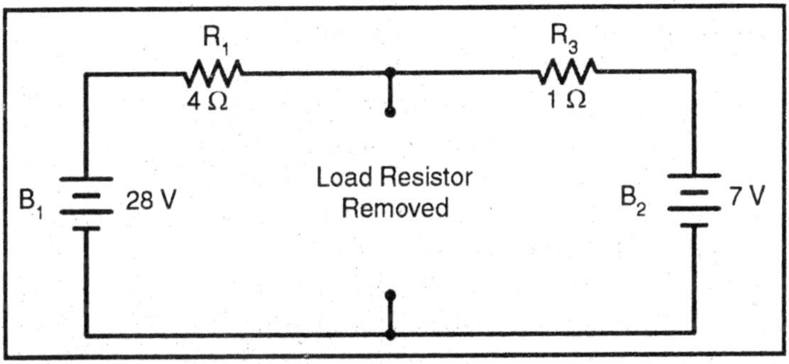

Then, to find the Norton current (for the current source in the Norton equivalent circuit), place a direct wire (short) connection between the load points and determine the resultant current. Note that this step is exactly opposite the respective step in Thevenin's Theorem, where we replaced the load resistor with a break (open circuit):

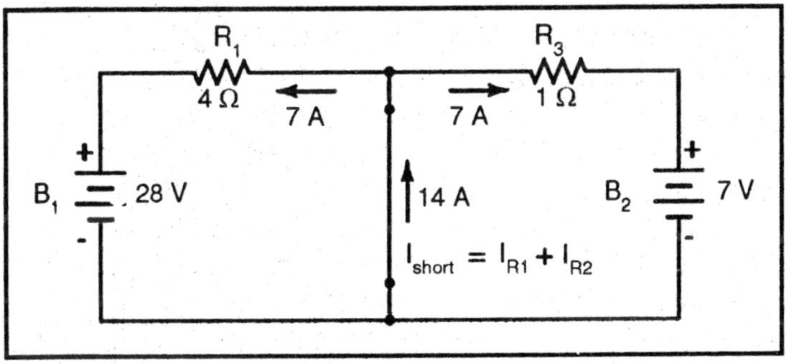

With zero voltage dropped between the load resistor

connection points, the current through R_1 is strictly a function of B_1's voltage and R_1's resistance: 7 amps (I=E/R). Likewise, the current through R_3 is now strictly a function of B_2's voltage and R_3's resistance: 7 amps (I=E/R). The total current through the short between the load connection points is the sum of these two currents: 7 amps + 7 amps = 14 amps.

This figure of 14 amps becomes the Norton source current (I_{Norton}) in our equivalent circuit:

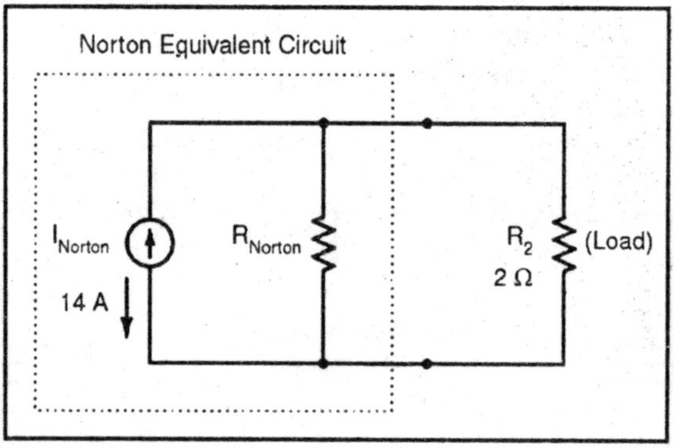

Remember, the arrow notation for a current source points in the direction opposite that of electron flow. Again, apologies for the confusion. For better or for worse, this is standard electronic symbol notation. To calculate the Norton resistance (R_{Norton}), we do the exact same thing as we did for calculating Thevenin resistance ($R_{Thevenin}$): take the original circuit (with the load resistor still removed), remove the power sources (in the same style as we did with the Superposition Theorem: voltage sources replaced with wires and current sources replaced with breaks), and figure total resistance from one load connection point to the other:

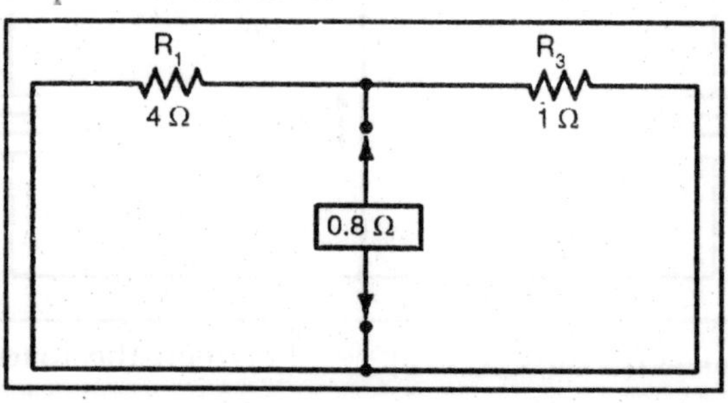

Now our Norton equivalent circuit looks like this:

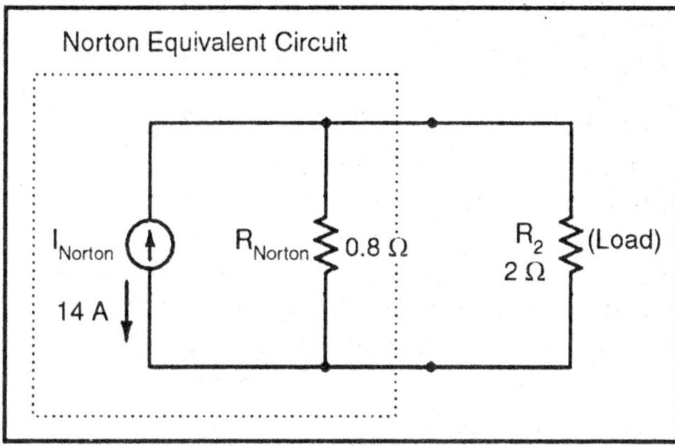

If we re-connect our original load resistance of 2 Ù, we can analyse the Norton circuit as a simple parallel arrangement:

	R_{Norton}	R_{Load}	Total	
E	8	8	8	Volts
I	10	4	14	Amps
R	0.8	2	571.43m	Ohms

As with the Thevenin equivalent circuit, the only useful information from this analysis is the voltage and current values for R_2; the rest of the information is irrelevant to the original circuit. However, the same advantages seen with Thevenin's Theorem apply to Norton's as well: if we wish to analyse load resistor voltage and current over several different values of load resistance, we can use the Norton equivalent circuit again and again, applying nothing more complex than simple parallel circuit analysis to determine what's happening with each trial load.

REVIEW

- Norton's Theorem is a way to reduce a network to an equivalent circuit composed of a single current source, parallel resistance, and parallel load.

Steps to follow for Norton's Theorem:

- Find the Norton source current by removing the load resistor from the original circuit and calculating current through a short (wire) jumping across the open connection points where the load resistor used to be.
- Find the Norton resistance by removing all power sources in the original circuit (voltage sources shorted and current sources open) and calculating total resistance between the open connection points.
- Draw the Norton equivalent circuit, with the Norton current source in parallel with the Norton resistance. The load resistor re-attaches between the two open points of the equivalent circuit.
- Analyse voltage and current for the load resistor following the rules for parallel circuits.

DESCRIPTION ABOUT NETWORK ANALYSIS

Generally speaking, *network analysis* is any structured technique used to mathematically analyse a circuit (a "network" of interconnected components). Quite often the technician or engineer will encounter circuits containing multiple sources of power or component configurations which defy simplification by series/parallel analysis techniques. In those cases, he or she will be forced to use other means. This chapter presents a few techniques useful in analyzing such complex circuits. To illustrate how even a simple circuit can defy analysis by breakdown into series and parallel portions, take start with this series-parallel circuit:

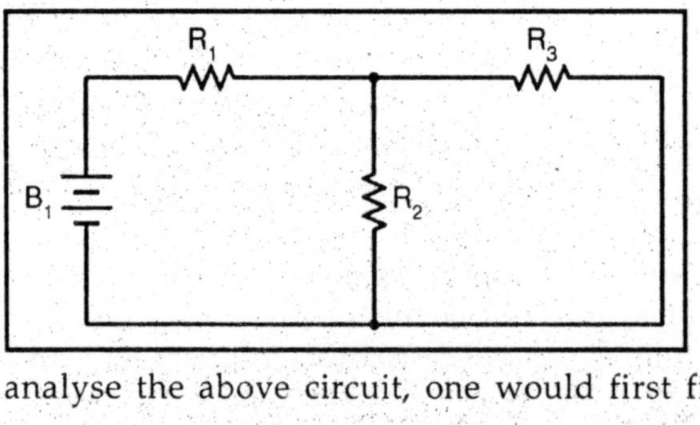

To analyse the above circuit, one would first find the

equivalent of R_2 and R_3 in parallel, then add R_1 in series to arrive at a total resistance. Then, taking the voltage of battery B_1 with that total circuit resistance, the total current could be calculated through the use of Ohm's Law (I=E/R), then that current figure used to calculate voltage drops in the circuit. All in all, a fairly simple procedure.

However, the addition of just one more battery could change all of tha0t:

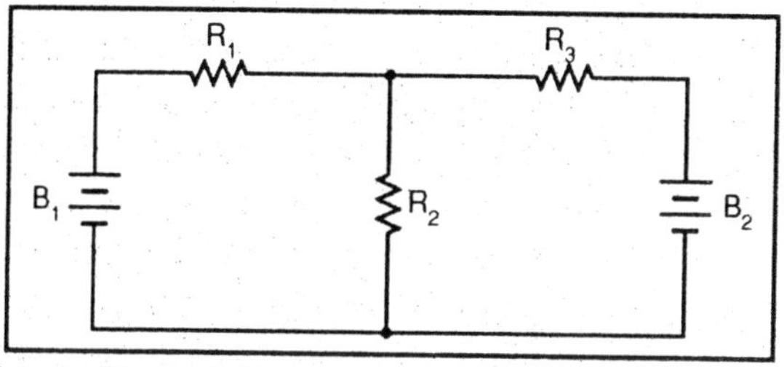

Resistors R_2 and R_3 are no longer in parallel with each other, because B_2 has been inserted into R_3's branch of the circuit. Upon closer inspection, it appears there are *no* two resistors in this circuit directly in series or parallel with each other. This is the crux of our problem: in series-parallel analysis, we started off by identifying sets of resistors that *were* directly in series or parallel with each other, reducing them to single equivalent resistances.

It should be clear that this seemingly simple circuit, with only three resistors, is impossible to reduce as a combination of simple series and simple parallel sections: it is something different altogether. However, this is not the only type of circuit defying series/parallel analysis:

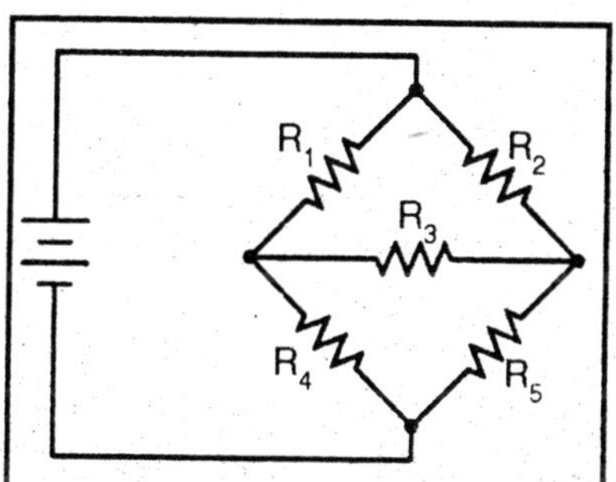

Here we have a bridge circuit, and for the sake of example we will suppose that it is *not* balanced (ratio R_1/R_4 not equal to ratio R_2/R_5).

If it were balanced, there would be zero current through R_3, and it could be approached as a series/parallel combination circuit ($R_1-R_4//R_2-R_5$). However, any current through R_3 makes a series/parallel analysis impossible. R_1 is not in series with R_4 because there's another path for electrons to flow through R_3. Neither is R_2 in series with R_5 for the same reason. Likewise, R_1 is not in parallel with R_2 because R_3 is separating their bottom leads. Neither is R_4 in parallel with R_5.

Although it might not be apparent at this point, the heart of the problem is the existence of multiple unknown quantities. At least in a series/parallel combination circuit, there was a way to find total resistance and total voltage, leaving total current as a single unknown value to calculate (and then that current was used to satisfy previously unknown variables in the reduction process until the entire circuit could be analysed). With these problems, more than one parametre (variable) is unknown at the most basic level of circuit simplification.

With the two-battery circuit, there is no way to arrive at a value for "total resistance," because there are *two* sources of power to provide voltage and current (we would need *two* "total" resistances in order to proceed with any Ohm's Law calculations). With the unbalanced bridge circuit, there is such a thing as total resistance across the one battery (paving the way for a calculation of total current), but that total current immediately splits up into unknown proportions at each end of the bridge, so no further Ohm's Law calculations for voltage (E=IR) can be carried out.

So what can we do when we're faced with multiple unknowns in a circuit? The answer is initially found in a mathematical process known as *simultaneous equations* or *systems of equations,* whereby multiple unknown variables are solved by relating them to each other in multiple equations. In a scenario with only one unknown (such as every Ohm's Law equation we've dealt with thus far), there only needs to be a single equation to solve for the single unknown:

E = 1R (E is unknown; 1 and R are known).

$I = \frac{E}{R}$ (I is unknown; E and R are known)

$R = \frac{E}{1}$ (R is unknown; E and 1 are known)

However, when we're solving for multiple unknown values, we need to have the same number of equations as we have unknowns in order to reach a solution. There are several methods of solving simultaneous equations, all rather intimidating and all too complex for explanation in this chapter. However, many scientific and programmable calculators are able to solve for simultaneous unknowns, so it is recommended to use such a calculator when first learning how to analyse these circuits.

Later on we'll see that some clever people have found tricks to avoid having to use simultaneous equations on these types of circuits.

REVIEW

- Some circuit configurations ("networks") cannot be solved by reduction according to series/parallel circuit rules, due to multiple unknown values.
- Mathematical techniques to solve for multiple unknowns (called "simultaneous equations" or "systems") can be applied to basic Laws of circuits to solve networks.

Chapter 2

Concept of Circuits

NETWORK

DRM IS PASSIVE AND ACTIVE

Passive DRM protects its content from onlookers who do not have a DRM-enabled client. Encryption is generally used for Passive DRM, so that the content is meaningless garbage unless you have the right bits in your client. I consider this 'passive' protection, because the data is inaccessible by default and only becomes accessible if you have the right kind of client, with the right key.

Active DRM, then, would be a scheme where protection is only provided if the client in use is one that is correctly coded to block access where it has not been specifically granted. This is a scheme in which the data is readily accessible to most normal viewers/players, but has a special code that tells a DRM-enabled viewer/player to hide the content from people who haven't been approved.

The whole problem is his two categories are a false distinction. You can't arbitrarily draw a line through a system and say "this is passive, this is active." For your CSS example, if you consider a given player's decryption code along with an arbitrary encrypted DVD, you have a system with both active and passive elements. If you leave out either of those elements, you have a disc that won't play or a player with no disc, the only perfectly secure system (assuming your cryptography is good.) When judging the efficiency of new compression schemes, the size of the decoder is added to the size of the compressed data to get a fair assessment of its efficiency. Otherwise you could win contests with a one-byte file and a 10 GB decoder programme that simply contains all the actual data.

Whichever way you design a system, complexity is being pushed from one party to another but never eliminated. For DVD, where most of the complexity is in the player, there is a huge variety of player implementations that each have their own bugs. The author of every disc needs to test against many combinations of players because of that problem.

Likewise, if you push the complexity onto the disc by including executable code there, the player gets simpler but the disc could be buggy. However, in that case, the content author will get a bad reputation for the buggy disc.

This doesn't just apply to DRM. While he might consider a MPEG4-AVC video file as "passive" in his terminology, it is really a complex series of instructions to the decoder. Look at the number of different but valid ways to encode video and you'll see it's closer to a programme than to "passive" data.

Now in his definition for "Active DRM", he is not actually describing the general class of software protection techniques. He is describing a system that is poorly-designed, often due to an attempt to retrofit DRM onto an existing system without it. Ofcourse it makes sense that if you have two ways to access the content, one with DRM and the other without, the additional complexity makes no sense to the end-user or mass copiers. It may make economic sense to the content author, but they have to weigh the potential risks to their business also (annoying users vs. stopping some casual copying.)

Even assuming his terminology makes sense, the Windows Media Centre system he references is actually a combination of "active" and "passive". The cable video stream is encrypted ("passive"), and the Windows DRM component is "active". In particular, it has a "black box" DLL that checks the host environment and hashes various items to derive a key, hence the problem.

All I can distill from what Alun says is "an unprotected system is made more complex by adding DRM." I agree, but this doesn't say anything larger about "active" versus "passive" DRM.

Full disclosure: I was previously one of the designers of the Blu-ray protection layer (BD+), a unique approach to disc protection that involves both cryptography and software protection. You can consider me biased, but my analysis should be able to stand on its own.

So, if you encrypt a file, and hand me the encrypted bits on a disk, that's what I would call "passive DRM", because its protection does not require me to be running the DRM client code. Only its unlocking requires me to run the DRM client code (or a reasonable facsimile thereof) and get a key.

That latter kind of DRM can only function in an environment where users can be prevented from _not_ running the DRM client code. This is why sometimes successful attempts by various content producers to pass legislation that prevents users from choosing what software to run – or not run – on their own machines.

Sure, you can say that there's a continuum between these two extremes – and that the fact that "Passive DRM" requires a key to be given to the client certainly makes it possible to hack – but I think the distinction is worth making.

So, now that I've further explained myself, do you agree with me that there is a valid distinction between the two types of protection? Are we simply quibbling about names?

- *Encrypted file + DRM client ("ProtectionX"):* DRM client attempts to verify environment before decrypting file. Assumes file is encrypted with standard symmetric crypto so that it can't be retrieved without at least one instance of DRM client.
- *Unprotected CD + autorun assassin software ("Assassin"):* CD is unprotected 16 bit samples readable by standard consumer grade equipment. In one specific environment and configuration (Windows with autorun), the assassin programme runs and attempts to interfere with any attempt to copy data.

What I'm trying to tell you is that your example of a file with a flag that says "do not copy" or a disc with autorun assassin software that is otherwise unprotected is a special case. It's an attempt to retrofit some kind of client-side control onto a format that doesn't already have it, not a type of pre-defined DRM.

No one, especially content producers, asks for an unprotected format and assumes they'll bolt on something later. All you're seeing is the fact that some formats made it out of the gate without protection (CD) or were weak and eventually broken (DVD-CSS). In those cases, some people make the effort to add optional "assassin" software, even if it only slows down some small per centage of users.

You and I both agree that this is ineffective, easily bypassed, and can interfere with legitimate use. But some companies have analysed the cost/benefit tradeoff and based on that information continue to deploy this approach rather than accept the alternative (do nothing).

Side note: The reason why this subject is particularly important to me is that some people are already using the concepts "active" and "passive" in terms of DRM. The claim is that crypto-based systems (*i.e.*, AACS) are "passive" and software protection (*i.e.*, BD+) are "active". This is nonsense since you have to consider the software that decrypts the disc part of the system, and thus there are no true "passive" systems.

Type 2 is no DRM at all, in essence. Whether it's a "broadcast flag" on television, or a CD with an executable rootkit, there are legitimate systems that won't even see that the content should be protected, and the temptation to deal with that is to somehow outlaw such systems.

It's effectively already happened – while multi-region DVD players are common and inexpensive in the UK, in the US, I find it difficult to find a player that is advertised as multi-region. While they're clearly not illegal, the marketplace here has somehow been coerced into making them unavailable.

Maybe my proposed two types should be "DRM" and "not DRM":) Well, it's some kind of restriction, although it's certainly not in the same class as something designed to be

DRM from the beginning. It's similar to people saying XOR with a constant is "encrypting the data" – nowhere near accurate. In that case, I call it "masking the data" since it is slightly transformed. So what's the term for "unprotected, but with a couple speedbumps bolted on the side"?

DUAL-FREQUENCY-BAND PATCH ANTENNA

A dual-frequency-band patch antenna capable of resonating at two different operating frequencies. The dual-frequency-band patch antenna includes two conductive plates that are placed side by side, with one conductive plate sized to resonate at a first frequency and the other conductive plate sized to resonate at a second frequency. The conductive plates are electrically joined by an electrical connection and share a single feed from a transceiver. The conductive plates are separated from a ground plane by a dielectric medium, which according to the present invention is very thin in terms of the wavelength at which the antenna operates ($i(*i.e.*), 'low-profile'). When the antenna is excited at the first frequency, the first conductive plate resonates like a conventional patch antenna. The second conductive plate (204), however, acts like a parasitic patch, thereby increasing the overall bandwidth and efficiency. Conversely, when the antenna is excited at the second frequency, the second conductive plate (204) resonates and the first conductive plates acts like a parasitic patch.

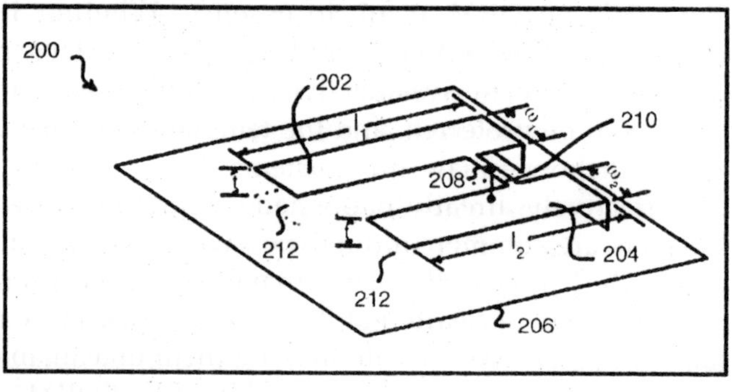

PASSIVE ELEMENTS

Most of the Circuit elements have at least two leads

(electrical terminals). They are characterized by voltage across the terminals and current flowing through the device, *see Figure* this is V-I characterization of device.

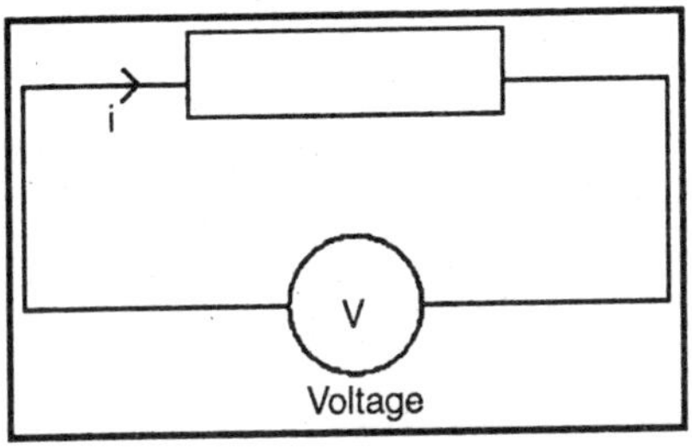

Resistance

Across this element, if we applied a voltage source and observe the current, then we will observer that $v \alpha i$, that is, $v = Ri$, where R is the resistance measured in Ohms Figure.

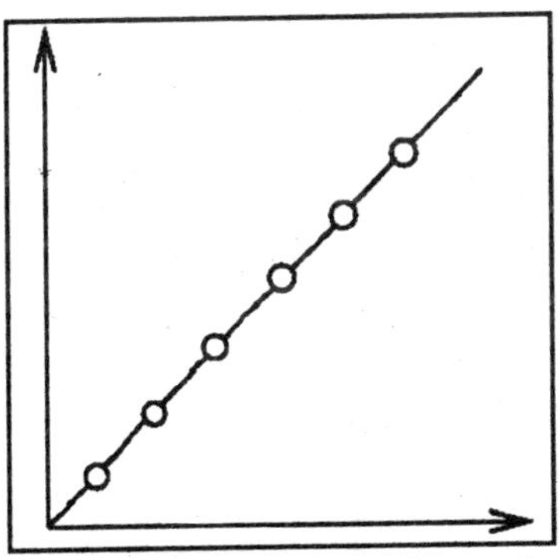

Higher the value of R, larger the voltage required to achieve the same current. Voltage proportional to current - is Ohm's law (It was deduced hueristically by experiments for metals, by George Simon Ohm). For lamp, this law is not true, as with increase in current, temperature of bulb increases, causing the increase in resistance. Hence Ohms law is not strictly true for lamp. But for most of the practical purposes and for this course the Ohm's law holds true for the resistive circuit elements.

The resistive elements (resistances) can be fixed or variable. Commonly use resistor types are carbon film and wirewound. The example of variable resistor is potentiometre.

For a material with length l and cross-sectional area A, the resistance will be,

$$R = \frac{P^l}{A},$$

where P is specific resistivity (property of material). Inverse of resistance is conductance.

$$G = \frac{1}{R},$$

measured in mhos or Seimens.

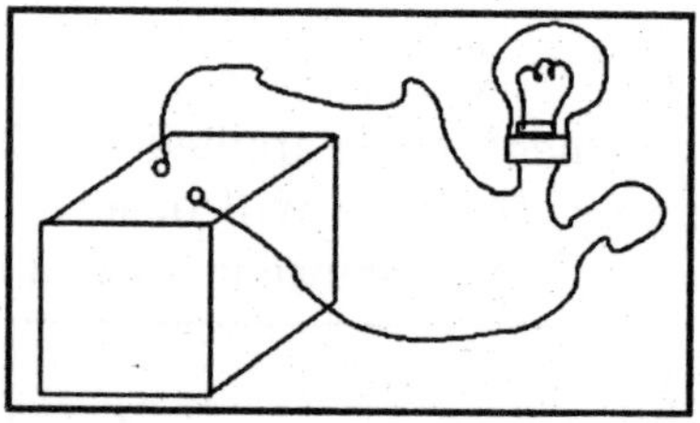

Fig. Bulb as a resistive load

In Fig. property of interest is resistance between A and B. It is dependent on details of wire, connector, filament, material, and shape. It can be abstracted as simple resistance Figure.

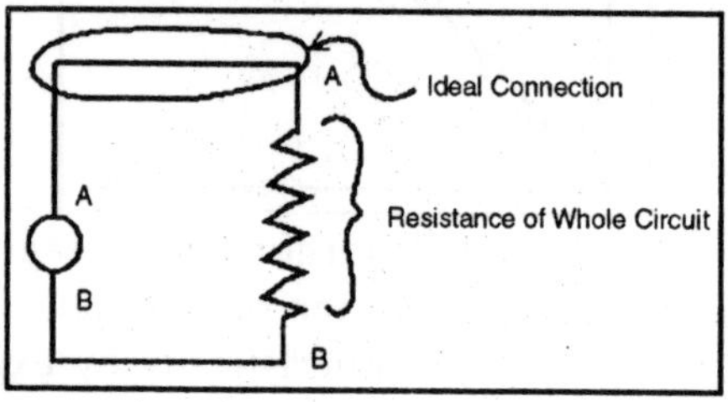

Fig. Bulb abstracted as a simple resistance

Inductance

It another important basic circuit element. Current flowing in a wire causes generation of magnetic field intensity *(H)*. *H* is independent of material medium surrounding the current carrying wire. The *H* leads to magnetic flux density.

$$B = \mu_r \mu_0 H.$$

Here $\mu 0$ is absolute permeability of vaccum. μr is relative permeability of material where B is measured. The flux flowing around wire links with the conducting wire. And if the flux linkage changes it lead to generation of EMF (electromotive force) which try to oppose the change in flux. This means it tries to nullify the change in current.

The current i causes production of magnetic flux,

$$E = \frac{d\phi}{dt}$$

is the EMF of a single turn. Here, $BA = f$.

Thus, the total EMF,

$$v = N\frac{d\phi}{dt},$$

where N is the number of turns.

Defining the inductance,

$$-v = L\frac{di}{dt}, L$$

is inductance and measured in Henry. See the output current of sinusoidal v applied across an inductor in Figure.

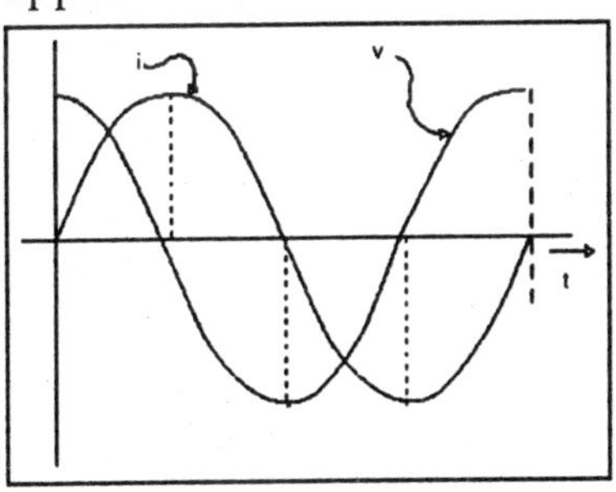

Fig. V and I as function of t for an Inductor

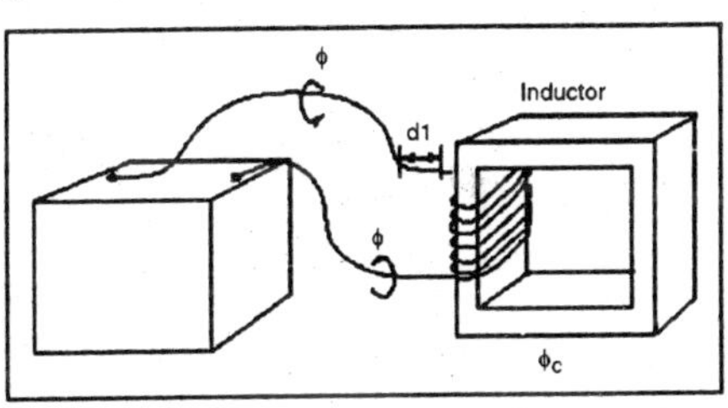

Fig. Physical Implementation of an Inductor

In lumped model, inductance is considered only due to element. The inductance due to wires connecting it to other elements is neglected.

Analysis of circuit:

To find voltage or currents in an element of interest. One can also find voltage and current in all the elements of circuit. Lumped simplified model of resitance, inductance and capacitance.

Capacitance: $i = c\frac{dv}{dt}$, c capacitance.

The magnitude of charge on either plate is given by $q = cv$.

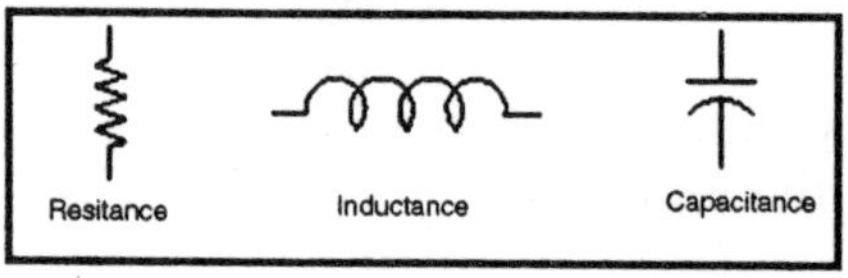

Fig. Symbols for Resitance, Inductance and Capacitance

Lumped Model

Shape, material, wire, connectors - effect of each is assumed to be due to single entity shown by the symbols in the diagram. In actual resistance, inductance and capacitance are distributed all across the circuits. For most practical purpose, lumped model- satisfactory.

Series and Parallel Connections

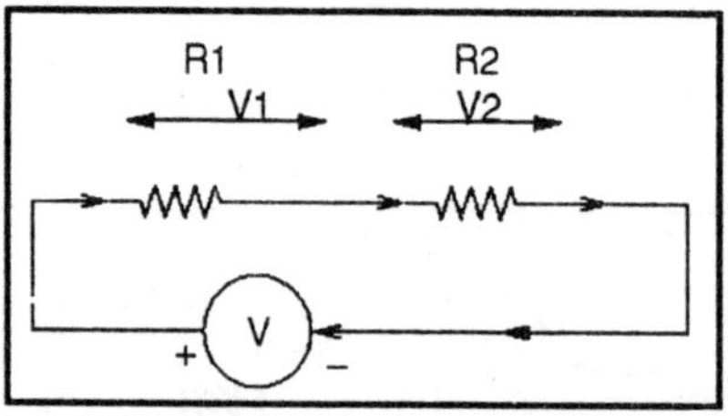

Fig. A Series Connection of Resis tors

Kirchoff's Volatage Law

In a circuit, if your start from a point A and tranverses

the circuit in any fashion and reaches back to point A, the total sum of potential changes should be zero. This has to be true since, same point cannot have two different potentials.

Kirchoff's Current Law

At any point in the network, total amount of current entering and leaving the point has to be equal to rate of accumulation of charge at the point. Since, in the circuits ordinarily the points where one circuit element is connected to other circuit element (these points are called nodes) do not store charge sum of incomming current has to be equal to sum of outgoing currents.

Thus, for series model, we get relations:

$$v_1 + v_2 = v$$

$$iR_1 + iR_2 = v$$

$$i(R_1 + iR_2) = v$$

Thus equivalent resistance of series connection is:

$$R = R_1 + R_2$$

For parallel connection, we get the relations:

$$v_1 = v_2 = v$$

$$i_1R_1 = i_2R_2 = v$$

$$i = i_1 + i_2 = \frac{v}{R_1} + \frac{v}{R_2}$$

$$i\frac{R_1R_2}{R_1 + R_2} = v$$

Thus, equivalent resistance here is:

$$R = \frac{R_1R_2}{R_1 + R_2}.$$

Similarly, we calculate equivalent inductance and capacitance for series and parallel cases.

Inductances in Series

$$v = v_1 + v_2 = L_1\frac{di}{dt} + L_2\frac{di}{dt}$$

$$v = (L_1 + L_2)\frac{di}{dt}$$

Thus, $L = L_1 + L_2$

Inductances in Parallel

$$v_1 = v_2 = v$$

$$v_1 = L_1 \frac{di_1}{dt}$$

$$v_2 = L_2 \frac{di_2}{dt}$$

$$\frac{v_1}{L_1} + \frac{v_2}{L_2} = v(\frac{1}{L_1} + \frac{1}{L_2}) = \frac{d(i_1 + i_2)}{dt} = \frac{di}{dt}$$

Thus, $L = \dfrac{L_1 L_2}{L_1 + L_2}$

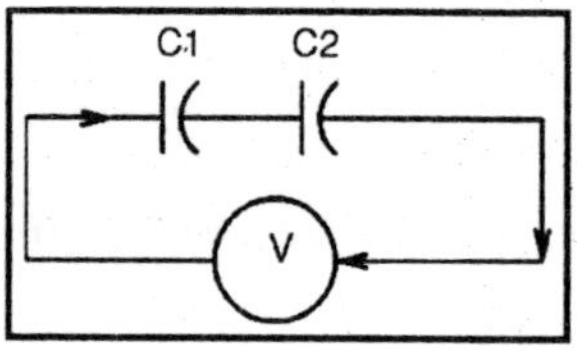

Fig. Series Connection of Capacitances

Capacitances in Series

$$i = C_1 \frac{dv_1}{dt} = C_2 \frac{dv_2}{dt}$$

$$i = (\frac{1}{C_1} + \frac{1}{C_2}) = \frac{d(v_1 + v_2)}{dt}$$

$$i = \frac{C_1 C_2}{C_1 + C_2} \frac{dv}{dt}$$

Thus, $C = \dfrac{C_1 C_2}{C_1 + C_2}$

Capacitances in parallel:

$$i_1 = C_1 \frac{dv}{dt}$$

$$i = i_1 + i_2 = (C_1 + C_2)\frac{dv}{dt}$$

Thus, $C = C_1 + C_2$

Inductance of a Solenoid (Coil)

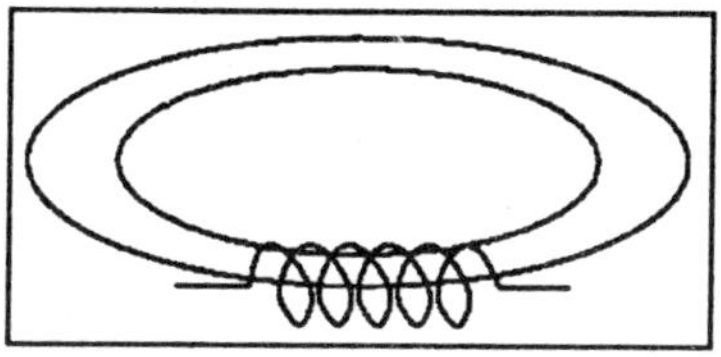

Fig. A Solenoid Diagram Showing Magnetic circuit path Lengths

Define l = magnetic circuit path length

A = magnetic circuit crossectional area.

Inductance: Assuming that H is same in the closed path of length l.

$$\oint \vec{H} \cdot \vec{dl} = Ni$$

$$H.l = Ni$$

$$H = \frac{Ni}{L}$$

$$B = \mu H$$

$$B = \frac{\mu Ni}{l}$$

Thus, $\phi = \dfrac{\mu NiA}{l}$

Now, $v = L\dfrac{di}{dt} = N\dfrac{d\phi}{dt}$

Thus, $v = \dfrac{\mu N^2 A}{l}\dfrac{di}{dt}$

Which gives $L = \dfrac{\mu N^2 A}{l}$

Here $H.l = Ni$ is magnetomotive force (equivalent of eletromotive force - EMF in magnetic domain), and flux ϕ is equivalent of current in magnetic domain. The Magnetic reluctance=

$$\frac{1}{\mu A}$$

is then equivalent of resistance in magentic domain.

Linearity: When elemental change in cause ΔC, always leads to same elemental change in effect *i.e.,* ΔE, then the system is said to be linear. In general, for a system let input $x(t)$ lead to output $y(t)$, and a small perturbation in $\delta x(t)$ causes a small perturbation $\delta y(t)$ in output. If perturbation $\delta x(t)$ alwasy leads to same perturbation of $\delta y(t)$ in the output irrespective of any $x(t)$, then the system is linear.

The implication of the above is that if input $x_1(t)$ causes output $y_1(t)$, and $x_2(t)$ causes $y_1(t)$, then $Ax_1(t)+Bx_2(t)$will cause an output of $Ay_1(t)+By_2(t)$ This is principal of superposition.

ACTIVE ELEMENTS

Active components are those that have gain or directionality. They include Semiconductors (Diode, Transistors, Operational Amplifier and other Solid State Devices) and Thermionic Valves (Vacuum Tubes).

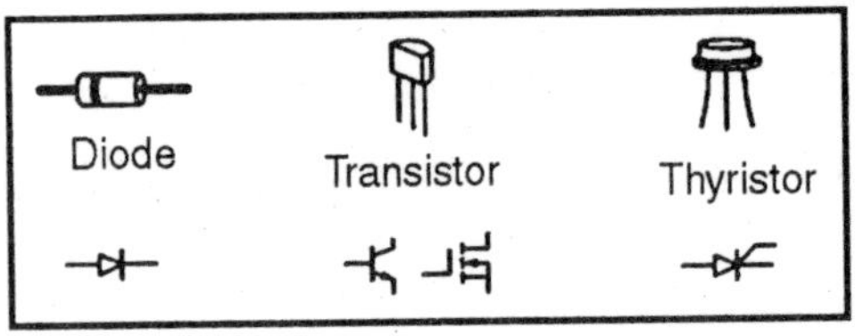

VOLTAGE AND CURRENT SOURCES

Real voltage sources can be represented as ideal voltage sources in series with a resistance r, the ideal voltage source having zero resistance. Real current sources can be represented as ideal current sources in parallel with a resistance r, the ideal current source having infinite resistance.

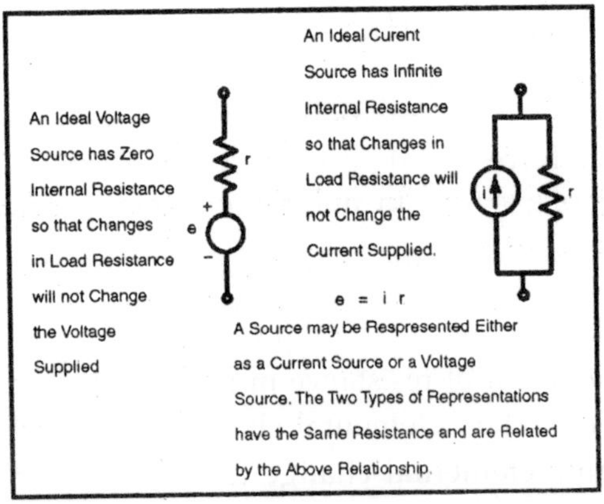

AN IDEAL VOLTAGE SOURCE

Looking at voltage sources first, let's consider the concept of an ideal voltage source. When you buy a battery you buy a nine (9) volt battery or a 1.5 v battery, or a twelve (12) volt car battery or some other battery that has a specified voltage. Clearly those batteries - electrical sources - are assumed to have a certain voltage - whatever it is - that doesn't change much.

When you buy a power supply for your calculator or a telephone answering machine or some other device you need to look at the voltage you need for the power supply. They usually come in a few specified voltages.

It seems clear that many sources are designed to give you some particular voltage and to attempt to maintain a constant voltage.

Below we'll consider simple models for voltage sources that maintain a constant voltage and we'll take a look at how you can represent that kind of device. A voltage source that can maintain a constant voltage - no matter what you do to it, like drawing a lot of current, or putting it in a situation where current flows through it - is an ideal voltage source.

That's what we are going to examine in this lesson. We'll leave it for another lesson to look at sources that are not ideal. So, let's answer that question above. What is an ideal voltage source? And there's another question implied. Why worry about an ideal voltage source since nothing like that exists in the real world. We'll take those questions in order.

CURRENT VOLTAGE SOURCES

The concept of an ideal voltage source is pretty simple, and it was really embedded in the previous discussion.

- An ideal voltage source is a voltage source that maintains the same voltage across the source's terminals no matter what current is drawn from the terminals of the source or what current flows into the terminals.
- That's it in a nutshell. If the source is a DC Source, we can plot a voltage current plot for an ideal voltage source. The plot is shown below. However, we need

to define terms. Here is a circuit symbol for an ideal voltage source.

In this symbol, we assume the following:

- The voltage across the terminals is denoted as V_t.
- The load current flowing from the source to a load (presumably a load is attached when the source is in a circuit) is denoted as I_L.
- With those definitions, here is the source symbol. It's just a circle with polarity indicated.

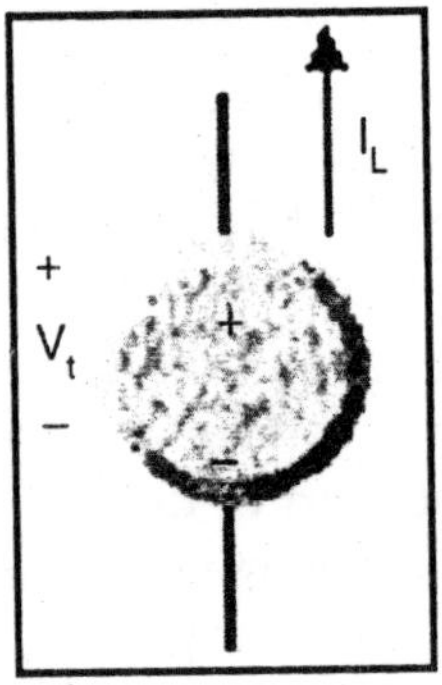

And, here is the plot of terminal voltage against load current.

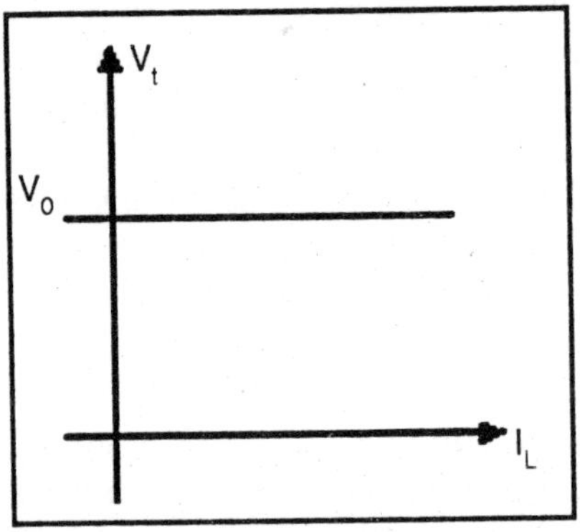

Given the discussion above, we can say:

- V_t = constant, no matter what the load current is. That's pretty much the description of the ideal voltage source. It's not too complex, but it is an important concept. In the next section we'll look at how you can put this concept to use. For the rest of this section we'll look at ideal current sources starting next.

IDEAL CURRENT SOURCE IN SIMPLE MODE

An ideal current source is a simple model for many current sources. It is reminiscent of the ideal voltage source but with voltage and current interchanged. Here is the story.

- There is a special circuit symbol for an ideal current source.

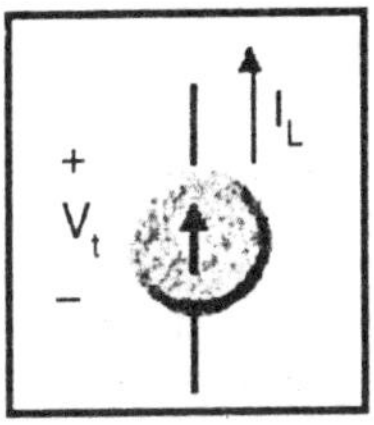

- I_L = constant, no matter what the terminal voltage is.
- The plot of load current against terminal voltage is similar to the plot for an ideal voltage source, but voltage and current are interchanged. Here is the plot.

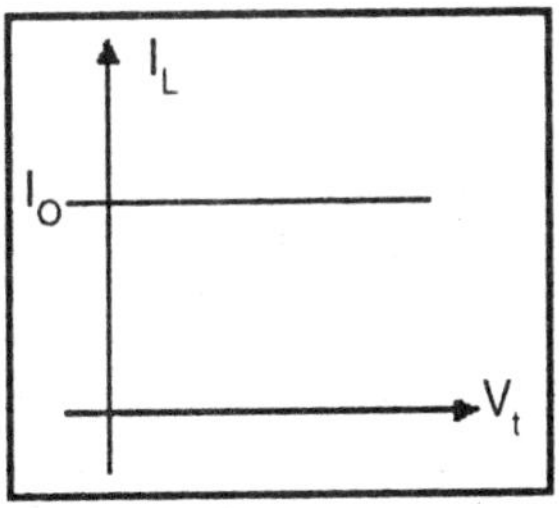

Notice that an ideal current source is somewhat similar to an ideal voltage source. However, when you use an ideal source - usually when doing circuit analysis- there is a significant difference in the analysis. However, that's getting ahead of the story. We first have to worry about how you would "use" an ideal source, when we know that there is no such thing as an ideal source, i.e. a source that is "perfect" in some way.

USING IDEAL SOURCES

The idea of using ideal sources is something that you may rebel at. After all, there is no such thing as an ideal source anywhere in the world. You can't pull an ideal source off the shelf in the lab, so why are we even talking about them? The answer to that question is that you use ideal sources when you have a non-ideal (a real source) source in a circuit. There are two important things to note.

- There are some sources that are very good sources and that can be modelled as ideal sources. (And when that happens, be grateful.) Some situations like that include the following.
  - A power supply in the lab. Many times you connect a power supply to some electronic circuit, for example, and when you connect the circuit you find that the output voltage from the power supply doesn't change measurably. (After all, power supply designers try to make that happen!) In that case, the power supply might be considered to be an ideal source at least as long as you are working on that particular circuit.
- There are many sources that do not perform ideally. However, it has proven to be possible to construct models of real sources, and those models often contain ideal source in combination with other ideal elements (like resistors, etc.).

  The vinin and Nortonequivalent circuits are examples of models of real sources that can account for loading effects *i.e.* drawing enough current from the source to change the output voltage) and they are widely used in circuit analysis. You will even find that manufacturers give you parametre values for Thevinin and Norton equivalents on the front panel of many instruments like function generators.

You often have situations in which the sources that you use can be approximated with ideal sources. Shown below is a bridge circuit powered by a battery. Often a battery maintains a pretty constant voltage across the terminals, so you may be able to replace the battery with an ideal voltage source when you analyse the circuit.

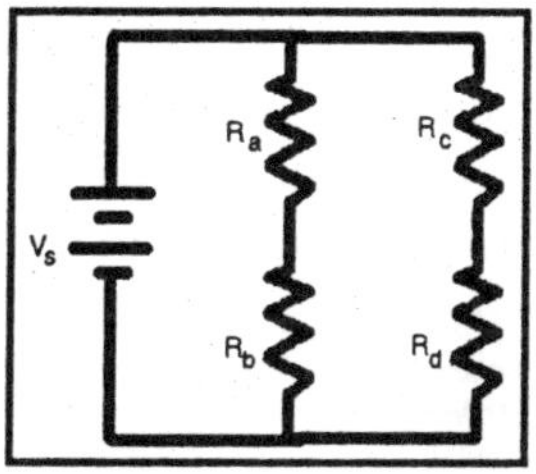

Here's the circuit with an ideal voltage source substituted for the battery. At this point, you may know how to do the analysis so you're ready to go. If you don't know how to do the analysis, you'll get there in these lessons. to go to the first lesson on circuit analysis. However, you may want to spend more time looking at sources, particular sources that are not ideal.

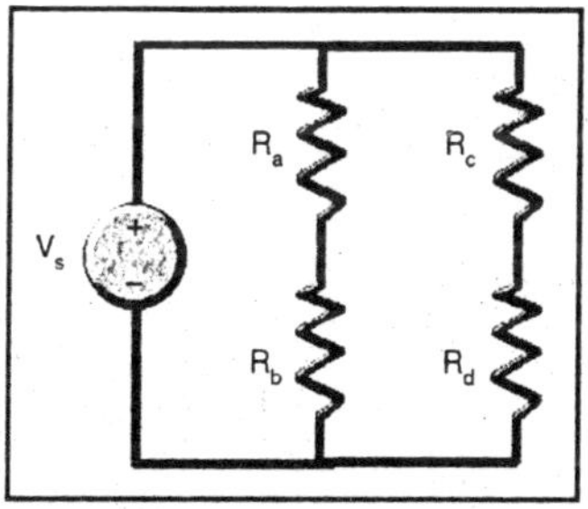

PC Controlled Voltage or Current Source

This project was begun as a means to charge and cycle NiCad batteries but has become a versatile tool for experiments requiring either a controlled voltage up to +30V and/or a current of +/- 2.5 Amps. If all you want to do is charge and recycle NiCad batteries, then, with hindsight, I'd advise you to buy one of the commerical units available. The circuit I will describe will do the job but it is experimental and has many short comings. It is more the beginning of a useful tool than one that has been fully developed. This circuit and the related software have taken me a good 2 years to reach even a marginal degree of reliability despite it's apparently simple actions.

My complete assembly uses the Discovery Series K-2805 Parrallel Port Interface Kit as distributed by Dick Smith Electronics as the source of control voltages and, if used as a charger, the Battery Cell measurements. Any other PC interface with at least 4 analogue inputs and 2 analogue outputs, both with 0 to +5V range, could be put to use.

Below is the circuit for one of the two identical voltage/ current sources included in my construction.

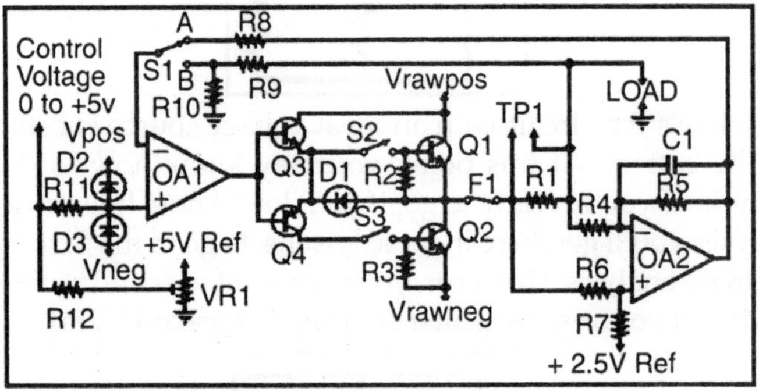

The components are:

Label	Component
OA1,OA2	2 parts of 4 OpAmp I.C. LM324
Q1,Q2	2N2055 (TO3 package NPN) or 2N3055
Q3	BC547 (any small NPN)
Q4	BC557 (any small PNP)
D1,D2,D3	signal diode rated over 30V
C1	1nF Green
F1	3A Fast Blow
R1	0.1R 5W
R2,R3	470R 1/2W
R4,R6	10K 1/4W
R5,R7	100K 1/4W
R8	47K 1/4W
R9	270K 1/4W
R10	47K 1/4W
R11	1K 1/4W
R12	100K 1/4W
VR1	500K

Vrawpos/neg is supplied in my construction by a seperate power supply module which comprises a centre tapped transformer, a 4amp rectifier and two 10000uF capacitors supplying unregulated 18/36V peak DC.

Typically I use this circuit with Prawpos at +18V and Prawneg and Ground connected to the centre tap. More typically Prawpos would be connected to +18V, Prawneg to -18V and Ground to the centre tap. For higher voltages, Ground can be connected to the -18V supply but great care must be taken this is not connecting a -18V voltage to the PC interface ground which may be earthed through the PC.

The load in my case is a NiCad battery of between 1.2 and 24V, but it could really be anything; active loads like batteries are just more of a challenge to control.

Note: to discharge a 1.2V battery will required a negative power source, but for larger voltages, Prawneg may be connected to ground as described above.

Q1 is sources voltage to the load and Q2 sinks voltage from the load. Naturally they get hot and so need good heat sinking and some thermal isolation from other components and sensitive fingers. I was lucky to find a preloved aluminium box with an internal frame on which I mounted the transistors. Air holes either end have further helped to keep the box reasonably cool.

Source or Sink actions can be disabled by opening S2 or S3 respectively. R2 and R3 then clamp the transistor base to prevent transistor operation. Even then, some minor leakage will be apparent through R2 & R3 themselves; a better solution may be to have S2 & S3 in circuit to the load but their rating would need to be increased accordingly.

Transistors Q3 & Q4 buffer the controlling action of opamp OA1 which, throught the magic of feedback, eliminates any dead zone due to the Vbe voltage drops in Q3, Q4 or Q1 (*i.e.* without feedback, any control voltage would first have to overcome these voltage drops before current could flow to or from the load).

The link between the base of Q3 and the emitter of Q4 is to enable lower discharge voltages to ground when Vrawneg is to ground (enabling discharge to +0.7V with S2 closed, +1.1V with S2 open). This will only work with D1 there to allow the emitter of Q4 to rise above ground. D1 is also there to prevent Base to Emitter breakdown of Q4 (which occurs when the output of OA1 exceeds +6V); it doesn't really prevent it, but it does limit the current from OA1s output through the base to the emitter. The same it true for Q3 if the output of OA1 goes below -6V. Trying to understand this apparently simple circuit has caused me a few headaches. As mentioned already, to reliably discharge a 1.2V cell a split supply is really necessary but it is a bit scary letting a PC connect a -15V to a +1.2V cell.

A voltage between 0 and +5V is fed to OA1 either from a PC interface or, when the interface is unplugged, from the manually operated VR1 (it assumes that the impedance of the Analogue Voltage from the PC interface is much lower than R12). I have used this circuit to charge batteries using only the manual contol but VR1 does not provide a stable control for low currents and frequent adjustment is necessary (not to mention a good alarm clock to remind you to turn it off). Still the manual control is particularly useful for quick diagnostics.

With SW1 in position A, Current Source Mode, negative feedback to OA1 comes from a current measurement taken across R1 via OA2. Since the OA2 is referenced to +2.5V, 0V on the control voltage should result in -2.5Amps (sinking) through R1 (assuming a suitable load is connected).

If the control voltage rises to +5.0V, then +2.5Amps (sourcing) through R1 should result. If the load imedance is too high, the output voltage at the load will rise or fall towards the supply rail voltages (more on this in a moment).

With SW1 in position B, Voltage Source Mode, negative feedback to OA1 comes from the output voltage, divided by a factor of 6 by R9 and R10. A control voltage of 0V will result in 0V at the output, +5V will (should) result in +30V (if Vpos is > +30V) at the output. I am thinking R10 could be connected to Vneg to allow negative outputs but it seemed more useful to have the output voltage controlled relative to ground.

For the purposes of Battery charging, SW1 remains in position A. With the opamp used (LM324) and the the supply conditioning I provide for the opamp (see below), the maximum supply voltage in current source mode is Vrawpos - 4.5V (13.5V for 18V supply). As such, running this system off a car battery will only allow you to charge a 5 cell NiCad. Using an OpAmp which can source and sink closer to supply voltage and possibly connecting the OpAmp supply directly to Vrawpos (quite okay for a 13.5V car battery supply) may allow you to charge 7 cell packs off a car battery but I haven't tried it. I think we'll leave fast charging off car batteries to a more dedicated unit. I originally thought it would be useful to measure both output voltage and output current and since both feedback signals should be in the 0 to 5V range, these can be measured with the PC Interface that I use which has 10 analogue inputs. Current measurement turns out to be of no particular use in battery charging as any errors in the control of the source or sink current arise from the +2.5V reference voltage which in turn are reflected in the measurement with a few extra errors thrown in.

The current measurement may be of more use where this device is being used as a voltage source or in analysing someone elses charger have done this to determine the parametres for an unlabelled NiCad pack that came with a toy car and plug pack charger. Current can also be measured by a sensitive voltmetre across R1 using test poit TP1.

Below is the circuit I used to supply the I.C.s and voltage references:

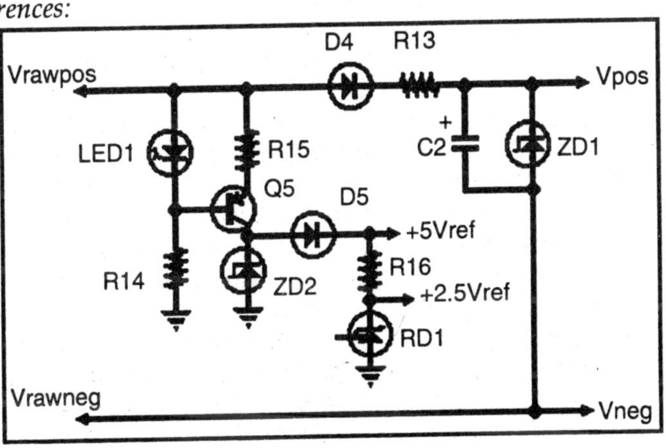

The components are:

Label	Component
D1	Rectifier Diode rated over 30V and 100
mA	
ZD1	Zener Diode 30V 1W
ZD2	Zener Diode 5.1V 400mW
Q5	BC557 PNP
RD1	LM336-2.5V Voltage Reference
D5	Small Rectifier Diode (~30mA duty)
LED1	Whatever
R13	330R 1W
R14	2.2K 1/4W
R15	110R 1/4W
R16	1K 1/4W
C2	10uF 50VW

The LM324 is happy with voltages up to 30V, so this supply is clamped at 30V with ZD1 and C2 removes a good deal of ripple from the unregulated supply. The value of R13 was found by trial and error to match the I.C. supply current. D4 is there to prevent the inevidable mistake.

The voltage reference circuit draws more curreThe LM324 is happy with voltages nt than I would have liked so I have used a current source from the unregulated supply to feed 5V Zener ZD2 which in turn feeds the 2.5V reference circuit. Originally I used a simple resistor to feed ZD2 but with the supply varying from 10 to 36V, the current to ZD2 also varied greatly and all sorts of problems followed. Q5 controls the current to approx. 20mA by using the voltage drop through LED1 as a reference voltage.

The diode is there to drop the "5V" level to about 4.7V. This is connected in my unit to the 5V supply in the PC interface (through a 1K resistor) so that both operate at the exact same level.

This was a good idea when the 2.5V reference was created by a simple divider, but now that I use a 2.5V reference diode, this link and diode D5 are nolonger necessary and I have only shown it as this is what is in my construction.

While it shows only one 2.5V ref. circuit (R16 & RD1), I have in my construction 2 circuits, one for each voltage current source circuit. This was necessary before I fitted Q5 to provide a constant current to ZD2. I suspect two 2.5V reference circuits are nolonger necessary but I haven't tried it yet.

One of the problems in using this Current Source for battery charging is that, until the software has taken control, this unit will initially attempt to sink 2.5Amps from the battery (control voltage at 0V). As such, either the battery must be left disconnected or S3 left open until the software has got itself together. If the computer crashes for some reason the results would be unpleasant. The simplest solution I have come up with so far is to disable the +2.5V reference circuit with the following alternative supply circuit.

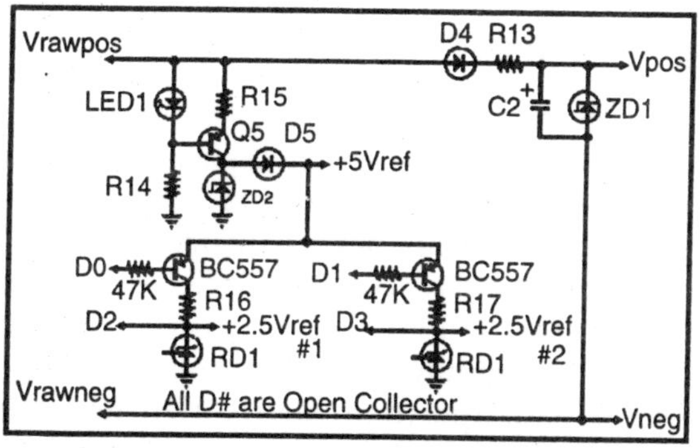

While D0 to D3 are open collector outputs from the PC Interface which must be enabled by the software before they sink any current. As such, while D0 and D1 are not enabled following boot up, the 2.5V reference will float between 0V (as will occur with zero control volts) and about 0.8V. On my unit at boot up, this results in about 100mA discharge of the connected battery. Not exactly passive, but not disasterous either.

As it happens, this modification provides additional versatility. By connecting D2 and D3 to the 2.5V reference voltages, these can be clamped to rise no greater than 0.9V. When this is so, 5V control voltage will now correspond to 5Amps source. Given the right power supply (mine could not handle this load for long) and the right load, this could be useful. There would be no affect in the voltage source mode.

This is the end of description of the circuit regarding voltage and current source control. The software (charger) I've written for an IBM PC compatible (based on Intel Microprocessor 8086 series and above) computer and the Discovery Series PC Interface, has a manual test mode and a battery charger mode.

Future modes intended for the voltage/current source will be added as they occur to me (or you). I will be considering release of the source code on a GNU licence with the hope that useful modification would be released back to me. In the mean while, if you would like the source code, my current email address is "rpact" with an ISP called "dingoblue.net.au".

LINEARITY AND LINEAR NETWORK

LINEARITY PROPERTY]

The growth in areas of application of electric circuits has led to an evolution from simple to complex circuits. To handle the complexity, engineers over the years have developed some theorems to simplify circuit analysis. Such theorems include Thevenin's and Norton's theorems. Since these theorems are applicable to linear circuits, we first discuss the concept of circuit linearity. In addition to circuit theorems, we discuss the concepts of superposition, source transformation, and maximum power transfer in this chapter. The concepts we develop are applied in the last section to source modeling and resistance measurement.

Linearity is the property of an element describing a linear relationship between cause and effect. Although the property applies to many circuit elements, we shall limit its applicability to resistors in this chapter. The property is a combination of both the homogeneity (scaling) property and the additive property. The homogeneity property requires that if the input (also called the excitation) is multiplied by a constant, then the output (also called response) is multiplied by the same constant. For a resistor, for example, Ohm's law relates the input I to the output v,

$$v=iR$$

If the current is increased by constant k, then the voltage increases correspondingly by k, that is,

$$kiR=kv$$

The additive property requires that the response to a sum of inputs is the sum of the responses to each input applied separately. Using the voltage current relationship of a resistor, if

$$v_1=i_1R$$

and

$$v_2=i_2R$$

then, applying (i_1+i_2) gives,

$$v=(i_1+i_2)R=i_1R+i_2R=v_1+v_2$$

We say that a resistor is a linear element because the voltage-current relationship satisfies both the homogeneity and the additive properties. In general, a circuit is linear if it is both additive and homogeneous. A linear circuit consists of only linear elements, linear dependent sources, and independent sources. A linear circuit is one whose output is linearly related (or directly proportional) to its input.note that since $p=i^2R=v^2/R$ (making it a quadratic function rather than a linear one), the relationship between power and voltage (or current) is nonlinear. Therefore, the theorems covered in this chapter are not applicable to power.

To illustrate the linearity principle, consider the linear circuit shown in Figure. The linear circuit has no independent sources inside it. It is excited by a voltage source vs, which serves as the input. The circuit is terminated by a load R. We may take the current i through R as the output. Suppose $v_S=10V$ gives i = 2 A. According to the linearity principle, $v_S=1V$ will give i = 0.2 A. By the same token, i = 1 mA must be due to $v_S=5mV$.

SUPERPOSITION

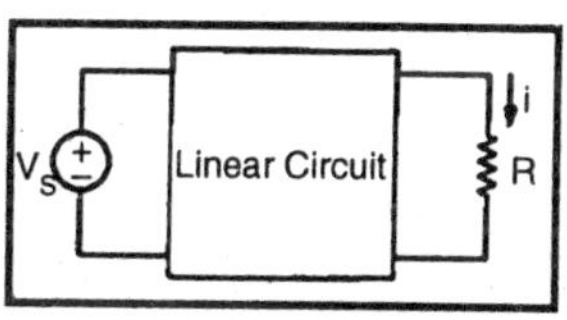

Fig. A linear Circuit with Input Vs and Output

If a circuit has two or more independent sources, one way to determine the value of a specific variable (voltage or current) is to use nodal or mesh analysis as. Another way is to determine the contribution of each independent source to the variable and then add them up. The latter approach is known as the superposition.

The idea of superposition rests on the linearity property. The superposition principle states that the voltage across (or current through) an element in a linear circuit is the algebraic sum of the voltages across (or currents through) that elements due to each independent source acting alone.

The principle of superposition helps us to analyse a linear circuit with more than one independent source by calculating the contribution of each independent source separately. However, to apply the superposition principle, we must keep two things in mind.

- We consider one independent source at a time while all other independent sources are turned off. This implies that we replace every voltage source by 0 V (or a short circuit), and every current source by 0 A (or an open circuit). This way we obtain a simpler and more manageable circuit.
- Dependent sources are left intact because they are controlled by circuit variables.

With this in mind, we apply the superposition principle in three steps:

Steps to apply superposition principle:

- Turn off all independent sources except one source. Find the output (voltage or current) due to that active source using the techniques covered inJ
- Repeat step 1 for each of the other independent sources.

Find the total contribution by adding algebraically all the contributions due to the independent sources.

Analyzing a circuit using superposition has one major disadvantage: it may very likely involve more work. If the circuit has three independent sources, we may have to analyse three simpler circuits each providing the contribution due to the respectively individual source. However, superposition does help reduce a complex circuit to simpler circuits through replacement of voltage sources by short circuits and of current sources by open circuits.

Keep in mind that superposition is based on linearity. For this reason, it not applicable to the effect on power due to each source, because the power absorbed by a resistor depends on the square of the voltage or current. If the power value is needed, the current through (or voltage across) the element must be calculated first using superposition.

SOURCE TRANSFORMATION

We have noticed that series-parallel combination and wye-delta transformation help simplify circuits. Source transformation is another tool for simplifying circuits. Basic to these tools is concept of equivalence. We recall that an equivalent circuit is one whose v-i characteristics are identical with the original circuit.

We saw that node-voltage (or mesh current) equations can be obtained by mere inspection of a circuit when the sources are all independent current (or all independent voltage) sources. It is therefore expedient in circuit analysis to be able to substitute a voltage source in series with a resistor for a current source in parallel with a resistor, or vice versa, as shown in Figure Either substitution is known as source transformation.

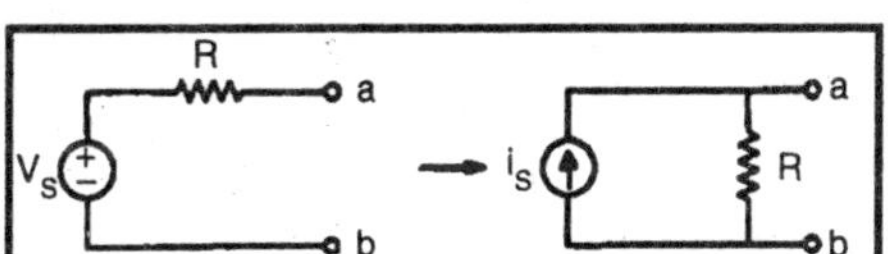

Fig. Transformation of Independent Sources.

A source transformation is the process of replacing a voltage source vs in series with a resistor R by a current source is in parallel with a resistor R, or vice versa.

The two circuits in Figure are equivalent – provided they have the same voltage–current relation at terminals a-b. It is easy to show that they are indeed equivalent. If the sources are turned off, the equivalent resistance at terminals a-b in both circuits is R. Also, when terminals a-b are short circuited, the short circuit current flowing from a to b is $i_{SG}=v_S/R$ in the circuit in the left hand side and $i_{SG}=i_S$ for the circuit on right hand side. Thus, $v_S/R=i_S$ in order for the two circuits to be equivalent.

Source transformation also applies to dependent sources, provided we carefully handle the dependent variable. As shown in Figure, a dependent voltage source in series with a resistor can be transformed to a dependent current source in parallel with the resistor or vice versa where we make sure that Equation is satisfied.

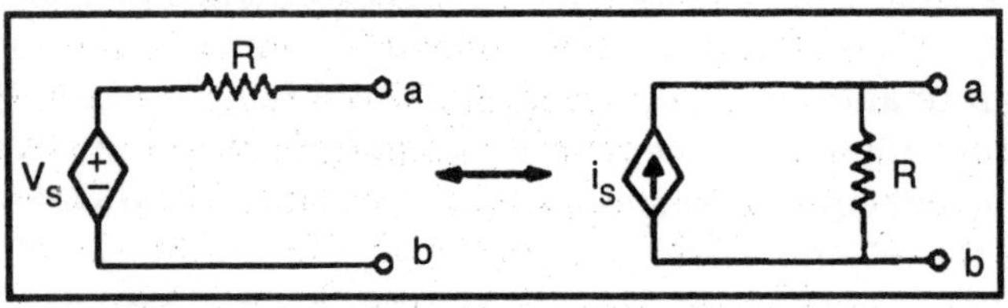

Fig. Transformation of Dependent Sources.

Like wye-delta transformation we studied in Chapter, a source transformation does not affect the remaining part of the circuit. When applicable, source transformation is a powerful tool that allows circuit manipulations to ease circuit analysis. However, we should keep the following points in mind when dealing with source transformation.

- Note from Figure the arrow of the current source is directed toward the positive terminal of the voltage source.
- Note from Equation that source transformation is not possible when R = 0, which is the case with an ideal voltage source. However, for a practical, nonideal voltage source, $R \neq 0$. Similarly, an ideal current source with $R = \infty$ cannot be replaced by a finite voltage source.

REQUIREMENTS FOR LINEARITY

A system is called linear, if it has two mathematical properties:homogeneity (hôma-gen-â-ity) and additivity. If you can show that a system has both properties, then you have proven that the system is linear. Likewise, if you can show that a system doesn't have one or both properties, you have proven that it isn't linear. A third property, shift invariance, is not a strict requirement for linearity, but it is a mandatory property for most DSP techniques.

When you see the term linear system used in DSP, you should assume it includes shift invariance unless you have reason to believe otherwise. These three properties form the mathematics of how linear system theory is defined and used. Later in this chapter we will look at more intuitive ways of understanding linearity. For now, let's go through these formal mathematical properties.

As illustrated in Figure; homogeneity means that a change in the input signal's amplitude results in a corresponding change in the output signal's amplitude. In mathematical terms, if an input signal of $x[n]$ results in an output signal of $y[n]$, an input of $kx[n]$ results in an output of $ky[n]$, for any input signal and constant, k.

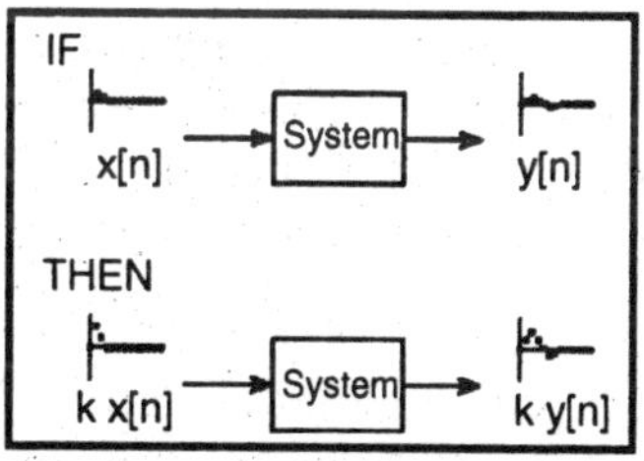

A simple resistor provides a good example of both homogenous and non-homogeneous systems. If, the input to the system is the voltage across the resistor, $v(t)$, and the output from the system is the current through the resistor, $i(t)$, the system is homogeneous.

Ohm's law guarantees this; if the voltage is increased or decreased, there will be a corresponding increase or decrease in the current.

Now, consider another system where the input signal is the voltage across the resistor, $v(t)$ but the output signal is the power being dissipated in the resistor, $p(t)$. Since power is proportional to the square of the voltage, if, the input signal is increased by a factor of two, the output signal is increase by a factor of four. This system is not homogeneous and therefore cannot be linear.

The property of additivity is illustrated in Figure; Consider a system where an input of $x_1[n]$ produces an output of $y_1[n]$. Further suppose that a different input, $x_2[n]$, produces another output, $y_2[n]$. The system is said to be additive, if an input of $x_1[n] + x_2[n]$ results in an output of $y_1[n] + y_2[n]$, for all possible input signals. In words, signals added at the input produce signals that are added at the output.

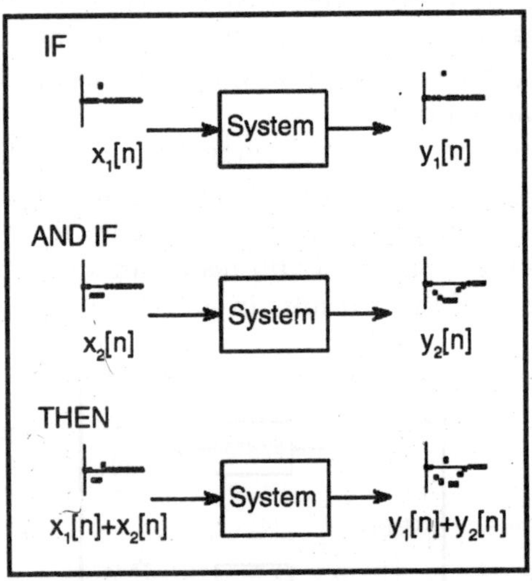

The important point is that added signals pass through the system without interacting. As an example, think about a telephone conversation with your Aunt Edna and Uncle Bernie. Aunt Edna begins a rather lengthy story about how well her radishes are doing this year. In the background, Uncle Bernie is yelling at the dog for having an accident in his favourite chair. The two voice signals are added and electronically transmitted through the telephone network. Since this system is additive, the sound you hear is the sum of the two voices as they would sound if transmitted individually. You hear Edna and Bernie, not the creature, Ednabernie.

A good example of a nonadditive circuit is the mixer stage

in a radio transmitter. Two signals are present: an audio signal that contains the voice or music, and a carrier wave that can propagate through space when applied to an antenna. The two signals are added and applied to a nonlinearity, such as a pn junction diode. This results in the signals merging to form a third signal, a modulated radio wave capable of carrying the information over great distances. As shown in Figure; shift invariance means that a shift in the input signal will result in nothing more than an identical shift in the output signal. In more formal terms, if an input signal of *x[n]* results in an output of *y[n]*, an input signal of *x[n + s]* results in an output of *y[n + s]*, for any input signal and any constant's.

Pay particular notice to how the mathematics of this shift is written, it will be used in upcoming chapters. By adding a constant's to the independent variable, *n*, the waveform can be advanced or retarded in the horizontal direction. For example, when $s = 2$, the signal is shifted left by two samples; when $s = -2$, the signal is shifted right by two samples.

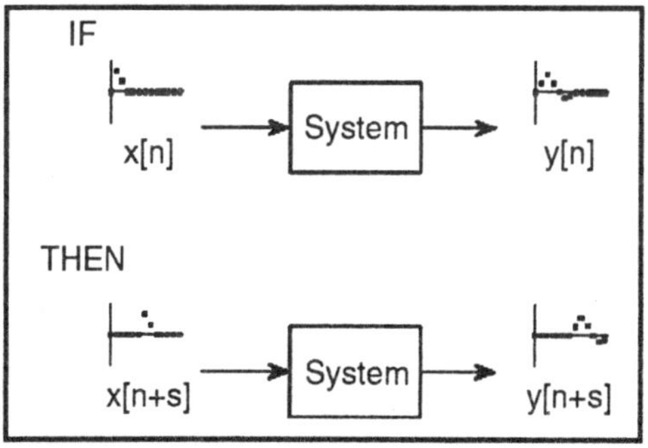

Shift invariance is important because it means the characteristics of the system do not change with time (or whatever the independent variable happens to be). If a blip in the input causes a blop in the output, you can be assured that another blipwill cause an identical blop. Most of the systems you encounter will be shift invariant. This is fortunate, because it is difficult to deal with systems that change their characteristics while in operation.

For example, imagine that you have designed a digital filter to compensate for the degrading effects of a telephone transmission line. Your filter makes the voices sound more natural and easier to understand. Much to your surprise, along comes winter and you find the characteristics of the telephone line have changed with temperature. Your compensation filter is now mismatched and doesn't work especially well. This situation may requirela more sophisticated algorithm that can adapt to changing conditions.

Why do homogeneity and additivity play a critical role in linearity, while shift invariance is something on the side? This is because linearity is a very broad concept, encompassing much more than just signals and systems. For example, consider a farm er selling oranges for $2 per crate and apples for $5 per crate. If the farmer sells only oranges, he will receive $20 for 10 crates, and $40 for 20 crates, making the exchange homogenous. If he sells 20 crates of oranges and 10 crates of apples, the farmer will receive: This is the same amount as if the two had been sold individually, making the transaction additive. Being both homogenous and additive, this sale of goods is a linear process. However, since there are no signals involved, this is not a system and shift invariance has no meaning. Shift invariance can be thought of as an additional aspect of linearity needed when signals and systems are involved.

LINEARITY AND SINUSOIDAL FIDELITY

Homogeneity, additivity, and shift invariance are important because they provide the mathematical basis for defining linear systems. Unfortunately, these properties alone don't provide most scientists and engineers with an intuitive feeling of what linear systems are about. The properties of static linearity and sinusoidal fidelityare often of help here. These are not especially important from a mathematical standpoint, but relate to how humans think about and understand linear systems. You should pay special attention to this section.

Static linearity defines how a linear system reacts when the signals aren't changing, *i.e.,* when they are *DC* or static. The output is the input multiplied by a constant. That is a graph of the possible input values plotted against the corresponding output values is a straight line that passes through the origin. This is shown in Figure for two common linear systems: Ohm's law for resistors and Hooke's law for springs. For comparison, Figure shows the static relationship for two nonlinear systems: a pn junction diode and the magnetic properties of iron.

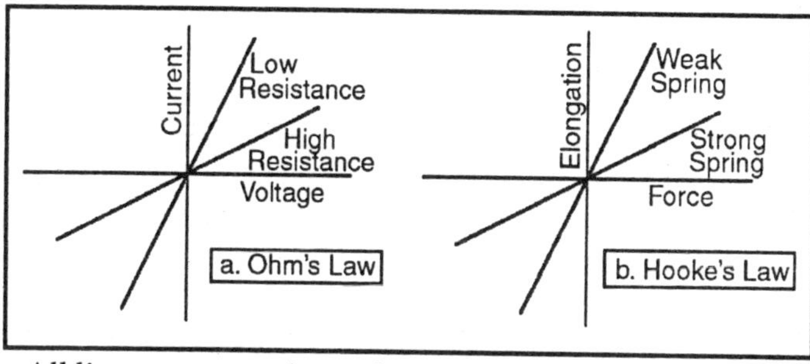

All linear systems have the property of static linearity. The opposite is usually true, but not always. There are systems that show static linearity, but are not linear with respect to changing signals. However, a very common class of systems can be completely understood with static linearity alone. In these systems it doesn't matter if the input signal is static or changing. These are called memoryless systems because the output depends only on the present state of the input, and not on its history.

For example, the instantaneous current in a resistor depends only on the instantaneous voltage across it, and not on how the signals came to be the value they are. If a system has static linearity and is memoryless then the system must be linear. This provides an important way to understand (and prove) the linearity of these simple systems.

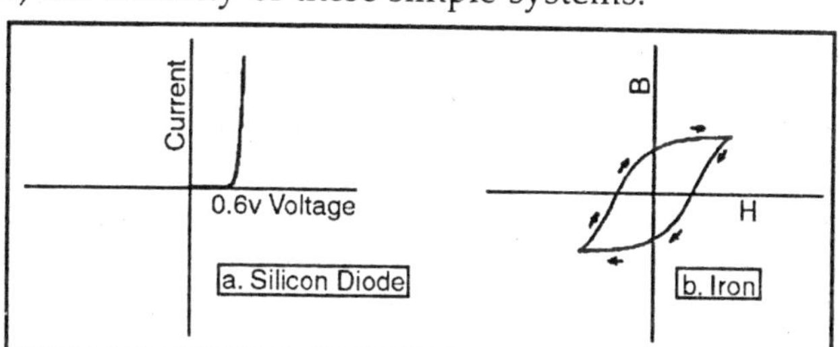

An important characteristic of linear systems is how they behave with sinusoids a property we will call sinusoidal fidelity: If the input to a linear system is a sinusoidal wave, the output will also be a sinusoidal wave, and at exactly the same frequency as the input. Sinusoids are the only waveform that have this property.

For instance, there is no reason to expect that a square wave entering a linear system will produce a square wave on the output. Although a sinusoid on the input guarantees a sinusoid on the output, the two may be different in amplitude and phase. This should be familiar from your knowledge of electronics: a circuit can be described by its frequency response, graphs of how the circuit's gain and phase vary with frequency.

If a system always produces a sinusoidal output in response to a sinusoidal input, is the system guaranteed to be linear? The answer is no, but the exceptions are rare and usually obvious. For example, imagine an evil demon hiding inside a system, with the goal of trying to mislead you. The demon has an oscilloscope to observe the input signal and a sine wave generator to produce an output signal. When you feed a sine wave into the input, the demon quickly measures the frequency and adjusts his signal generator to produce a corresponding output.

Of course, this system is not linear because it is not additive. To show this, place the sum of two sine waves into the system. The demon can only respond with a single sine wave for the output. This example is not as contrived as you might think phase lock loops operate in much this way.

To get a better feeling for linearity, think about a technician trying to determine if an electronic device is linear. The technician would attach a sine wave generator to the input of the device and an oscilloscope to the output. With a sine wave input, the technician would look to see if the output is also a sine wave.

For example, the output cannot be clipped on the top or bottom, the top half cannot look different from the bottom half, there must be no distortion where the signal crosses zero, etc.

Next, the technician would vary the amplitude of the input and observe the effect on the output signal. If the system is linear, the amplitude of the output must track the amplitude of the input. Lastly, the technician would vary the input signal's frequency, and verify that the output signal's frequency changes accordingly.

As the frequency is changed, there will likely be amplitude and phase changes seen in the output, but these are perfectly permissible in a linear system. At some frequencies, the output may even be zero, that is, a sinusoid with zero amplitude. If the technician sees all these things, he will conclude that the system is linear. While this conclusion is not a rigorous mathematical proof, the level of confidence is justifiably high.

UNILATERAL AND BILATERAL ELEMENTS

BILATERAL AND UNILATERAL CONTRACTS

Remember, an offer plus an acceptance equals a contract.

In effect, the offeror (person making an offer) is making a promise to do something or to refrain from doing something IF the other person (offeree) will do something or refrain from doing something.

A little thought shows that an offer may be accepted either by a promise or by action. Where the terms of an offer permit, the contract is created by a promise in exchange for the offeror's problem. It is called a bilateral contract and both parties are bound as soon as the mutual promises are exchanged. For example, if A says "I'll sell you my car for \$400." B can accept by promising, "Okay, I'll buy your car for \$400." The contract for sale is immediately formed, and both parties are bound to perform as mutually promised. Notice that in the example just given, nobody mentioned when the actual exchange of car and money would take place.

If, having exchanged a promise for a promise, the buyer for days keeps putting off producing the money, the seller (regretting that it didn't occur to him to put a time limit on the transaction) will at some point become justified in deciding that the buyer has broken the agreement and that he (the seller) can sell the car to someone else.

If the disappointed buyer sues the seller, the courts will probably solve the problem with two of their favourite words: "reasonable" and "imply". The courts will most likely say that because there was no mention of time, that a "reasonable" time was implied by the contract, and that the seller was justified in selling the car to someone else after a week.

In other situations a judge might say that failure to mention time omitted an essential term from a contract and that the contract lacks specificity to the point that it is unenforceable meaning that the disappointed party just has to lump it. A contract formed by a promise for a promise is a bilateral contract. A unilateral contract is one in which the offer cannot be accepted by a promise only by action.

Where the terms of an offer show that it can be accepted only by an action "I'll pay you $200 if, you'll paint my barn by Tuesday" then there is no contract until the act has been performed.

Notice that the offer did not say, "I'll pay you $200 if you'll promise to paint my barn by Tuesday." A promise to paint the barn is not an acceptance and does not bind either party to anything. Only by the complete painting of the barn is the contract formed and the $200 owed. And if the offeror comes when the barn is almost painted and says, "Sorry, I withdraw my offer," there is no contract. The painter can, however, recover the value of his work under the doctrine of quasi-contract, to prevent the unjust enrichment of the offeror.

The difference between unilateral and bilateral contracts may seem academic and difficult, but it does have practical consequences.

One more point about acceptance: A person cannot put another in the position of accepting an offer by silence. A writes to B: "I will assume that you have accepted my offer to print your brochures for $500 unless I hear from you to the contrary by Tuesday." B is under no obligation to reply and his silence past Tuesday does not accept the offer.

ELEMENTS OF A CONTRACT

The requisites for formation of a legal contract are an

offer, an acceptance, competent parties who have the legal capacity to contract lawful subject matter, mutuality of agreement, consideration, mutuality of obligation and, if required under the statute of frauds, a writing.

Offer

An offer is a promise that is by its terms, conditional upon an act, forbearance or return promise being given in exchange for the promise or its performance. It is a demonstration of willingness to enter into a bargain, made so that another party is justified in understanding that his or her assent to the bargain is invited and will conclude it. Any offer must consist of a statement of present intent to enter a contract; a definite proposal that is certain in its terms and communication of the offer to the identified prospective offeree. If any of these elements are missing, there is no offer to form the basis of a contract.

Invitations to Bid

Preliminary negotiations are clearly distinguished from offers because they contain no demonstration of present intent to form contractual relations. No contract is formed when prospective purchasers respond to such terms, as they are merely invitations or requests for an offer. Unless this interpretation is employed, any person in a position similar to a seller who advertises goods in any medium would be liable for numerous contracts when there is usually a limited quantity of merchandise for sale.

An advertisement, price quotation or catalogue is customarily viewed as only an invitation to a customer to make an offer and not as an offer itself. The courts reason that an establishment might not have sufficient stock to satisfy potential demand and that it would not be reasonable for a customer to expect to form a binding contract by responding to advertisements that are intended to make consumers aware of a product for sale.

In addition, the courts have held that an advertisement is an offer for a unilateral contract that can be revoked at the will of the offeror, the business enterprise, prior to performance of its terms.

An exception exists, however, to the general rule on advertisements. When the quantity offered for sale is specified and contains words of promise, such as "first come, first served," courts enforce the contract where the store refuses to sell the product when the price is tendered. Where the offer is clear, definite, and explicit and no matters remain open for negotiation, acceptance of it completes the contract. New conditions may not be imposed on the offer after it has been accepted by the performance of its terms.

An advertisement or request for bids for the sale of particular property or the erection or construction of a particular structure is merely an invitation for offers that cannot be accepted by any particular bid. A submitted bid is however, an offer which upon acceptance by the offeree becomes a valid contract.

Mistake in Sending Offer

If an intermediary, such as a telegraph company, errs in the transmission of an offer, most courts hold that the party who selected that method of communication is bound by the terms of the erroneous message. The same rule applies to acceptances. In reaching this result, courts regard the telegraph company as the agent of the party who selected it. Other courts justify the rule on business convenience. A few courts rule that if there is an error in transmission, there is no contract, on the grounds that either the telegraph company is an independent contractor and not the sender's agent, or there has been no meeting of the minds of the parties. However, an offeree who knows, or should know, of the mistake in the transmission of an offer may not take advantage of the known mistake by accepting the offer; he or she will be bound by the original terms of the offer.

Termination of An Offer

An offer remains open until the expiration of its specified time period or, if there is no time limit, until a reasonable time has elapsed. A reasonable time is determined according to what a reasonable person would consider sufficient time to accept the offer.

The death or insanity of either party, before an acceptance is communicated, causes an offer to expire. If the offer has been accepted, the contract is binding, even if one of the parties dies there after. The destruction of the subject matter of the contract; conditions that render the contract impossible to perform; or the supervening illegality of the proposed contract results in the termination of the offer.

When the offeror, either verbally or by conduct, clearly demonstrates that the offer is no longer open, the offer is considered revoked when learned by the offeree. Where an offer is made to the general public, it can be revoked by furnishing public notice of its termination in the same way in which the offer was publicized.

Irrevocable Offers

An option is a right that is purchased by a person in order to have an offer remain open at agreed- upon price and terms, for a specified time, during which it is irrevocable. It constitutes an exception to the general rule that an offer may be withdrawn prior to acceptance. The offeror may not withdraw this offer because that party is bound by the consideration given by the offeree. The offeree is free, however, to decide whether or not to accept the offer.

Most courts hold that an offer for a unilateral contract becomes irrevocable as soon as the offeree starts to perform the requested act because that action serves as consideration to prevent revocation of the offer. Where it is doubtful whether the offer invites an act (as in the case of a unilateral contract) or a promise (as in the case of a bilateral contract), the presumption is in favour of a promise, and therefore a bilateral contract arises.

If an offer to form a unilateral contract requires several acts, it is interpreted as inviting acceptance by completion of the initial act. Performance of the balance constitutes a condition to the offeror's duty of performance. Where such an offer invites only a single act, it includes by implication a subsidiary promise to keep the offer open if the offeree will commence performance. Some courts hold that an offer for a unilateral contract may be revoked at any time prior to completion of the act bargained for, even after the offeree has partially performed it.

Rejection of An Offer

An offer is rejected when the offeror is justified in understanding from the words or conduct of the offeree that he or she intends not to accept the offer, or to take it under further advisement. Rejection might come in the form of an express refusal to accept an offer by a counteroffer, which is a new proposal that rejects the offer by implication; or by a conditional acceptance that operates as a counteroffer. The offer may continue, however, if the offeree expressly states that the counteroffer shall not constitute a rejection of the offer.

If an offer is rejected, the party who made the original offer no longer has any liability for that offer. The party who rejected the offer may not subsequently, at his or her own option, convert the same offer into a contract by a subsequent acceptance. In such a case, the consent of the offeror must be obtained for a contract to be formed.

Acceptance

Acceptance of an offer is an expression of assent to its terms. It must be made by the offeree in a manner requested or authorized by the offeror. An acceptance is valid only if the offeree knows of the offer; the offeree manifests an intention to accept; the acceptance is unequivocal and unconditional and the acceptance is manifested according to the terms of the offer.

The determination of a valid acceptance is governed by whether a promise or an act by the offeree was the bargained for response. Since the acceptance of a unilateral contract requires an act rather than a promise, it is unnecessary to furnish notice of intended performance unless the offeror requested it. If, however, the offeree has reason to believe that the offeror will not learn of the acceptance with reasonable promptness, the duty of the offeror is discharged unless the offeree makes a reasonable attempt to give notice; the offeror learns of the performance; or the offer indicates that no notice is required.

In bilateral contracts, the offer is effective when the offeree receives it. The offeree may accept it until the offeree receives notice of revocation from the offeror. Thereafter, an offer is revoked. Under the majority rule, which is known as the "mailbox rule," an acceptance is effective upon dispatch if the offeror explicitly authorizes that method of acceptance to be employed by the offeree, even if the acceptance is lost or destroyed in transit.

The majority rule is inapplicable, however, unless the acceptance is properly addressed and postage prepaid. It has no application to most option contracts, as acceptance of an option contract is effective only when received by the offeror.

If the acceptance mode used by the offeree is implicitly authorized by the offeror, such as the selection by the offeree of the same method used by the offeror, who neglected to designate a method of communication, an acceptance is effective upon dispatch if it is correctly addressed and the expense of its conveyance is prepaid. As with expressly authorized methods, the acceptance need not ever reach the offeror in order to form the contract.

In some jurisdictions, the use of a method not expressly or impliedly authorized by the offeror, even if more rapid in nature, results in a contract only upon receipt of the acceptance. In most jurisdictions, however, if the acceptance mode is inherently faster, it is deemed to be an impliedly authorized means, and acceptance is effective upon dispatch.

If the acceptance is transmitted by an expressly or impliedly authorized method to the wrong address, it is effective only upon receipt by the offeror. A wrong address is any address other than that implicitly authorized, even if the offeror were in a position to receive the acceptance at the substituted address.

An offeror who specifically states that there is no contract until the acceptance is received is entitled to insist upon the condition of receipt or upon any other provision concerning the manner and time of acceptance specified.

Rejection of the offer or revocation of conditional acceptance is effective upon receipt. A late or defective acceptance is treated as a counter offer, which will not result in a contract unless the offeror accepts it. If offers cross in the mail, there will be no binding contract, as an offer may not be accepted if there is no knowledge of it.

As a general rule, an offer may be accepted only by the offeree or an authorized agent. If, however, the offer is contained in an option contract, it may be the subject of an assignment or transfer without the consent of the offeror, unless the option involves a purchase on credit or expressly prohibits an assignment.

In contracts that do not involve the sale of goods, acceptance must comply exactly with the requirements of the offer (this is known as the "mirror-image rule") and must omit nothing from the promise or performance requested. An offer of a prize in a contest, *for example,* becomes a binding contract when a contestant successfully complies with the terms of the offer. If a response to an offer purports to accept it, but adds qualifications or conditions, then it is a counteroffer and not an acceptance.

Acceptance may be inferred from the offeree's acts, conduct, or silence; but as a general rule, silence, without more, can never constitute acceptance. The effect of silence accompanied by ambiguity must be ascertained from all the circumstances in the case.

Prior dealings between the parties may create a duty to act. Silence or the failure to take some action under such circumstances might constitute acceptance. For example, if the parties have engaged in a series of business transactions involving the mailing of goods and payment by the recipient, the recipient will not be permitted to retain an article without paying for it within a reasonable time, due to their prior dealings.

A recipient who does not intend to accept the goods is

under a duty to inform the sender. Silence, where there is a duty to speak, prevents the offeree from rejecting an offer and the offeror from claiming that there is no acceptance. If ownership rights are exercised over an item, this might be deemed an acceptance.

Unsolicited Goods

At common law, the recipient of unsolicited goods in the mail was not required to accept or to return them, but if the goods were used, a contract and a concomitant obligation to pay for them were created. Today, in order to offer protection against unwanted solicitations, some state statutes have modified the common-law rule by providing that where unsolicited merchandise is received as part of an offer to sell, the goods are an out-right gift. The recipient may use the goods and is under no duty to return or pay for them unless he or she knows that they were sent by mistake.

Agreements to Agree

An "agreement to agree" is not a contract. This type of agreement is frequently employed in industries that require long-term contracts in order to ensure a constant source of supplies and outlet of production. Mutual manifestations of assent that are, in themselves, sufficient to form a binding contract are not deprived of operative effect by the mere fact that the parties agree to prepare a written reproduction of their agreement. In determining whether, on a given set of facts, there is merely an "agreement to agree" or a sufficiently binding contract, the courts apply certain rules. If the parties express their intention—either to be bound or not bound until a written document is prepared—then that intention controls. If they have not expressed their intention but they exchange promises of a definite performance and agree upon all essential terms, then the parties have formed a contract even though the written document is never signed.

If the expressions of intention are incomplete—as, for example, if a material term such as quantity has been left to further negotiation—the parties do not have a contract. The designation of the material term for further negotiation is interpreted as demonstrating the intention of the parties not to be bound until a complete agreement has been reached.

Competent Parties A natural person who agrees to a transaction has complete legal capacity to become liable for duties under the contract unless he or she is an infant, insane or intoxicated.

Infants An infant is defined as a person under the age of 18 or 21, depending on the particular jurisdiction. A contract made by an infant is voidable but is valid and enforceable until or unless he or she disaffirms it. He or she may avoid the legal duty to perform the terms of the contract without any liability for breach of contract. Infants are treated in such a way because public policy deems it desirable to protect the immature and naive infant from liability for unfair contracts that he or she is too inexperienced to negotiate on equal terms with the other party.

Once an infant attains majority (*i.e.*, the age at which a person is no longer legally considered an infant), he or she must choose either to disaffirm or avoid the contract, or to ratify or accept it. After reaching the age of majority, a person implicitly ratifies and becomes bound to perform the contract if he or she fails to disaffirm it within a reasonable time, which is determined by the circumstances of the particular case. A person who disaffirms a contract must return any benefits or consideration received under it that he or she still possesses. If such benefits have been squandered or destroyed, the person usually has no legal obligation to recompense the other party.

The law imposes liability on the infant in certain cases, however. Although the contract of an infant or other person may be voidable, the person still may be liable in quasi-contract in order to prevent unjust enrichment for the reasonable value of goods or services furnished if they are necessaries that are reasonably required for the person's health, comfort or education.

The majority of courts hold that an infant who willfully misrepresents, his or her age may never the less exercise the power to avoid the contract. As a general rule, however, the infant must place the adult party in the status quo ante (*i.e.*, his or her position prior to the contract).

The jurisdictions are in disagreement in regard to whether an infant is liable in tort (*i.e.*, a civil wrong other than breach of contract) for willful misrepresentation of his or her age. This divergence arises from the rule that a tort action may not be maintained against an infant if it essentially entails the enforcement of a contract. Some courts regard the action for fraud that would be commenced against the infant as being based on the contract.

Others rule that the tort is sufficiently independent of the contract so that the granting of relief would not involve indirect enforcement of the contract. The other party, however, is able to avoid a contract entered into on the basis of an infant's fraudulent misrepresentation with respect to age or other material facts because he or she is the innocent victim of the infant's fraud.

Mental Incapacity

When a party does not comprehend the nature and consequences of the contract when it is formed, he or she is regarded as having mental incapacity. A distinction must be drawn between those persons who have been adjudicated incompetent by a court and have had a guardian appointed and those mentally incompetent persons who have not been so adjudicated. A person who has been declared incompetent in a court proceeding lacks the legal capacity to enter into a contract with another.

Such a person is unable to consent to the contract, as the court has determined that he or she does not understand the obligations and effects of the contract. A contract made by such a person is void and without any legal effect. Neither party may be legally compelled to perform or comply with the terms of the contract. If there has been no adjudication of insanity, a contract made by a mentally incapacitated individual is voidable by him or her.

Many contract principles that apply to minors also apply to insane persons. There is an obligation to recompense the injured party where a voidable contract is avoided, and to pay for necessaries based upon quasi-contract for the reasonable value of the goods or services. The incompetent, a guardian or a personal representative after death may avoid the contract.

The incompetent may ratify a voidable contract only if they recover the capacity to contract. The right to avoid the contract belongs to the incompetent; the other party may not avoid the contractual obligation. A contract that is ordinarily voidable may not be set aside when it is inherently fair to both parties and has been executed to such an extent that the other party cannot be restored to the position that they occupied prior to the contract.

Intoxicated Persons

A contract made by an intoxicated person is voidable. When a person is inebriated at the time of entering into a contract with another and subsequently becomes sobre and either promises to perform the contract or fails to disaffirm it within a reasonable time after becoming sobre, then that person has ratified his or her voidable contract and is legally bound to perform.

Subject Matter Any undertaking may be the subject of a contract, provided that it is not proscribed by law. When a contract is formed in restraint of trade, courts will not enforce it because it imposes an illegal and unreasonable burden on commerce by hindering competition. Contracts that provide for the commission of a crime or any illegal objective are also void. Future rights and liabilities—performing or refraining from some designated act, or assuming particular risks or obligations may constitute the basis of a contract. An idea that never assumes concrete form at the time of disclosure, such as a concept for a short story, even though new and unusual, may not however, be the subject of a contract.

A person may not legally contract concerning a right that he or she does not have. A seller of a home who does not possess clear title to the property may not promise to convey it without encumbrances.

Mutual Agreement

There must be an agreement between the parties, or mutual assent, for a contract to be formed. In order for an agreement to exist, the parties must have a common intention or a meeting of minds on the terms of the contract and must subscribe to the same bargain.

The parties may settle one term at a time but their contract becomes complete only when they assent to the final term. An agreement is binding if the parties concur with respect to the essential terms and intend the agreement to be binding, even though all of the details are not definitely fixed. The quantity of goods are usually essential terms of the contract that must be agreed upon if the contract is to be enforced.

Exceptions to the rule requiring the terms of an agreement to be definite and certain are contained in article of the UCC, which permits the courts to imply reasonably the missing terms if the essential terms unambiguously demonstrate the mutual agreement of the parties.

Consideration

Consideration is a legal detriment that is suffered by the promise and that is requested by the promisor in exchange for his or her promise. A valid contract requires some exchange of consideration. As a general rule, in a bilateral contract, one promise is valid consideration for the other. In a unilateral contract, the agreed performance by the offeree furnishes the necessary consideration and also operates as an acceptance of the offer.

Consideration may consist of a promise; an act other than a promise; a forbearance from suing on a claim that is the subject of an honest and reasonable dispute; or the creation, modification, or destruction of a legal relationship.

It signifies that the promise will relinquish some legal right in the present, or that he or she will restrict his or her legal freedom of action in the future as an inducement for the promise of the other party. It is not substantially concerned with the benefit that accrues to the promisor.

Love and affection are not permissible forms of

consideration. A promise to make a gift contains no consideration because it does not entail a legal benefit received by the promisor or a legal detriment suffered by the promisee. Promissory estoppel is a doctrine by which a court enforces a promise that the promisor reasonably expects will induce action or forbearance on the part of a promisee, who justifiably relied on the promise and suffered a substantial detriment as a result.

Where a court enforces a promise by applying this doctrine, promissory estoppel serves as a substitute for the required consideration.

At common law, courts refused to inquire into the adequacy or fairness of a bargain, finding that the payment of some price constituted legally sufficient consideration. If one is seeking to prove mistake, misrepresentation, fraud, or duress or to assert a similar defence—the inadequacy of the price paid for the promise might represent significant evidence for such defenses, but the law does not require adequacy of consideration in order to find an enforceable contract.

Mutuality of Obligation

Where promises constitute the consideration in a bilateral contract, they must be mutually binding. This concept is known as mutuality of obligation. If one party's promise does not actually bind him or hers to some performance or forbearance, it is an illusory promise and there is no enforceable contract.

Where the contract provides one party with the right to cancel, there might be no consideration because of lack of mutuality of obligation. If there is an absolute and unlimited right to cancel the obligation, the promise by the party with the right of cancellation is illusory and the lack of consideration means that there is no contract.

If the power to cancel the contract is restricted in any manner, the contract is usually considered to be binding. Performance of a void promise in a defective bilateral contract may render the other promise legally binding, however. For example, in virtually all states, an oral contract to transfer title to land is not merely unenforceable, it is absolutely void. A seller who orally promises to transfer land to a purchaser, for which the purchaser orally promises a designated sum, may sue the purchaser for the price if the purchaser receives title to the land from the seller. The purchaser is not relieved of his or her promise to pay, because of the performance of the void oral promise by the seller.

A promise to perform an act that one is legally bound to do does not qualify as consideration for another promise. Past consideration consists of actions that occurred prior to the making of the contractual promise, without any purpose of inducing a promise in exchange. It is not valid, because it is not furnished as the bargained-for exchange of the present promise.

There are exceptions to this rule, such as a present promise to pay a debt that has been discharged in bankruptcy, which constitutes valid consideration because it renews a former promise to pay a debt that was supported by consideration.

Most states do not recognize moral obligation as consideration, as there is no acceptable method of setting the parametres of moral duty. Some courts will enforce a moral obligation where there has been a benefit conferred on the promisor.

Statute of Frauds

The statute of frauds was enacted by the English Parliament in 1677 and has since been the law in both England and in the United States in varying forms. It requires that certain types of contracts be in writing.

The principal characteristic of various state laws modeled after the original statute is the provision that no suit or action shall be maintained on a contract unless there is a note or memorandum of its subject matter, terms and conditions, and the identity of the parties, signed by the party to be charged or obligated under it or an authorized agent.

The purpose of the statute is to prevent the proof of a nonexistent agreement through fraud or perjury in actions for breach of an alleged contract.

R L AND C AS LINEAR ELEMENTS

L/R AND NOT LR

It is often perplexing to new students of electronics why the time-constant calculation for an inductive circuit is different from that of a capacitive circuit. For a resistor-capacitor circuit, the time constant (in seconds) is calculated from the product (multiplication) of resistance in ohms and capacitance in farads: T=RC. However, for a resistor- inductor circuit, the time constant is calculated from the quotient (division) of inductance in henrys over the resistance in ohms: T=L/R. This difference in calculation has a profound impact on the qualitative analysis of transient circuit response. Resistor-capacitor circuits respond quicker with low resistance and slower with high resistance; resistor-inductor circuits are just the opposite, responding quicker with high resistance and slower with low resistance. While capacitive circuits seem to present no intuitive trouble for the new student, inductive circuits tend to make less sense.

Key to the understanding of transient circuits is a firm grasp on the concept of energy transfer and the electrical nature of it. Both capacitors and inductors have the ability to store quantities of energy, the capacitor storing energy in the medium of an electric field and the inductor storing energy in the medium of a magnetic field. A capacitor's electrostatic energy storage manifests itself in the tendency to maintain a constant voltage across the terminals. An inductor's electromagnetic energy storage manifests itself in the tendency to maintain a constant current through it.

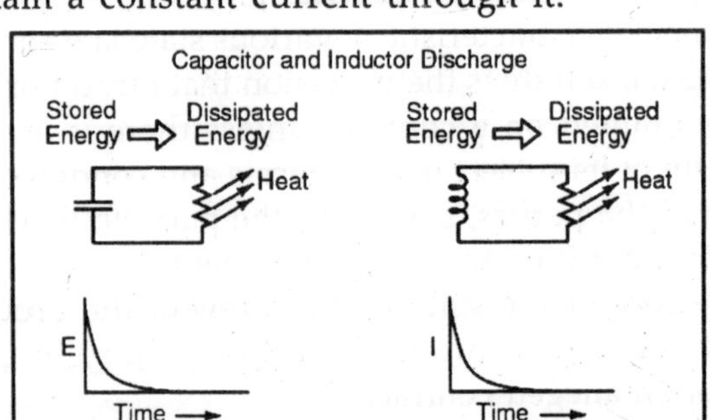

Let's consider what happens to each of these reactive components in a condition of discharge: that is, when energy is being released from the capacitor or inductor to be dissipated in the form of heat by a resistor:

In either case, heat dissipated by the resistor constitutes energy leaving the circuit, and as a consequence the reactive component loses its store of energy over time, resulting in a measurable decrease of either voltage (capacitor) or current (inductor) expressed on the graph.

The more power dissipated by the resistor, the faster this discharging action will occur, because power is by definition the rate of energy transfer over time.

Therefore, a transient circuit's time constant will be dependent upon the resistance of the circuit. Of course, it is also dependent upon the size (storage capacity) of the reactive component, but since the relationship of resistance to time constant is the issue of this section, we'll focus on the effects of resistance alone. A circuit's time constant will be less (faster discharging rate) if the resistance value is such that it maximizes power dissipation (rate of energy transfer into heat). For a capacitive circuit where stored energy manifests itself in the form of a voltage, this means the resistor must have a low resistance value so as to maximize current for any given amount of voltage (given voltage times high current equals high power). For an inductive circuit where stored energy manifests itself in the form of a current, this means the resistor must have a high resistance value so as to maximize voltage drop for any given amount of current (given current times high voltage equals high power).

This may be analogously understood by considering capacitive and inductive energy storage in mechanical terms. Capacitors, storing energy electrostatically, are reservoirs of potential energy. Inductors, storing energyelectromagnetically are (electrodynamically), are reservoirsof kinetic energy. In mechanical terms, potential energy can be illustrated by a suspended mass, while kinetic energy can be illustrated by a moving mass. Consider the following illustration as an analogy of a capacitor:

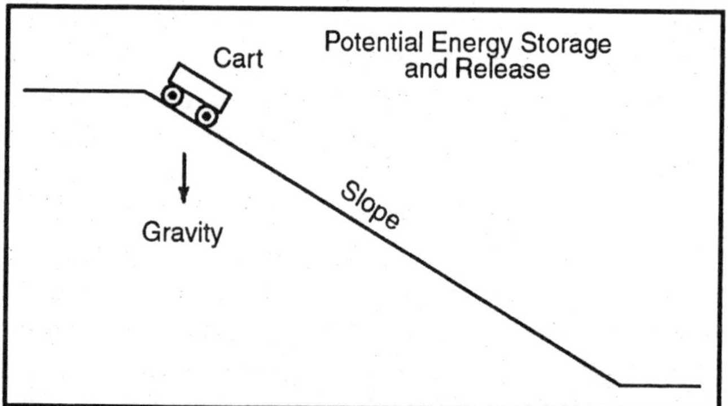

The cart, sitting at the top of a slope, possesses potential energy due to the influence of gravity and its elevated position on the hill. If we consider the cart's braking system to be analogous to the resistance of the system and the cart itself to be the capacitor, what resistance value would facilitate rapid release of that potential energy?

Minimum resistance (no brakes) would diminish the cart's altitude quickest, of course! Without any braking action, the cart will freely roll downhil, thus expending that potential energy as it loses height. With maximum braking action (brakes firmly set), the cart will refuse to roll (or it will roll very slowly) and it will hold its potential energy for a long period of time. Likewise, a capacitive circuit will discharge rapidly if its resistance is low and discharge slowly if its resistance is high.

Now let's consider a mechanical analogy for an inductor showing its stored energy in kinetic form:

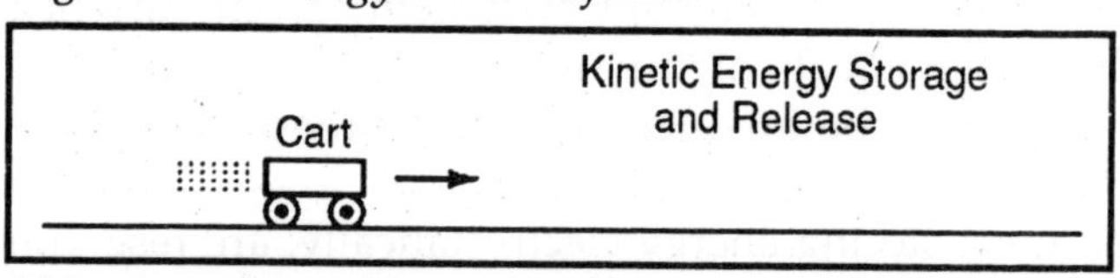

This time the cart is on level ground, already moving. Its energy is kinetic (motion), not potential (height). Once again if we consider the cart's braking system to be analogous to circuit resistance and the cart itself to be the inductor, what resistance value would facilitate rapid release of that kinetic energy?

Maximum resistance (maximum braking action) would slow it down quickest, of course! With maximum braking action, the cart will quickly grind to a halt, thus expending its kinetic energy as it slows down. Without any braking action, the cart will be free to roll on indefinitely (barring any other sources of friction like aerodynamic drag and rolling resistance), and it will hold its kinetic energy for a long period of time. Likewise, an inductive circuit will discharge rapidly if its resistance is high and discharge slowly if its resistance is low. Hopefully this explanation sheds more light on the subject of time constants and resistance, and why the relationship between the two is opposite for capacitive and inductive circuits.

SOURCE TRANSFORMATION

An ac voltage source V in series with an impedance Z can be replaced with an ac current source of value I =V/Z in parallel with the impedance Z. An ac current source I in parallel with an impedance Z can be replaced with an ac voltage source of value V = IZ in series with the impedance Z. Likewise, a dc voltage source *V* in series with a resistor *R* can be replaced with a dc current source of value $i = v/R$ in parallel with the resistor *R*; and vice versa.

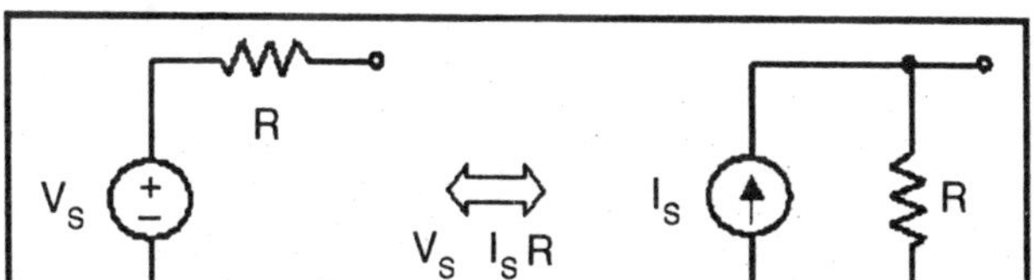

Source transformation is another tool for simplifying circuits. Basic to these tools is concept of equivalence. We recall that an equivalent circuit is one whose v-i characteristics are identical with the original circuit.

KIRCHHOFF'S LAWS

KIRCHHOFF'S CURRENT LAW (KCL)

KCL states that the algebraic sum of the currents in all

the branches which converge in a common node is equal to zero

$$\Sigma I_{tv}\square = \Sigma I_{out}$$

Kirchhoff's Voltage Law

Kirchhoff's Voltage Law states that the algebraic sum of the voltages between successive nodes in a closed path in the network is equal to xero,

$$\Sigma E = \Sigma IP$$

Soluton ustng Ktrchhoff,s Voltage anx current laws

Steps to Solve Circuit by Kirchhoff's Laws

- Construct circuit with circuit magic schematics editor

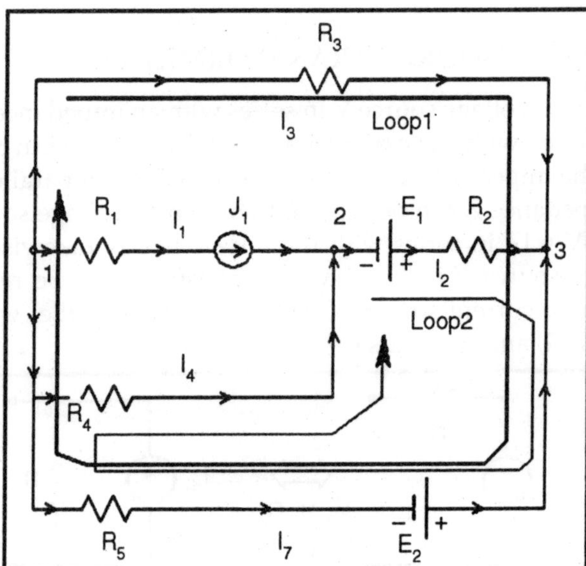

Circuit sample from circuit magic.

- Construct loops. Number of loops (and number of Kirhhoff's Voltage laws equations) can be determined using following formula. Loop can not include branches with current sources. Due current sources resistance equal infinity.

Let's take another look at our example series circuit, this time numbering the points in the circuit for voltage reference:

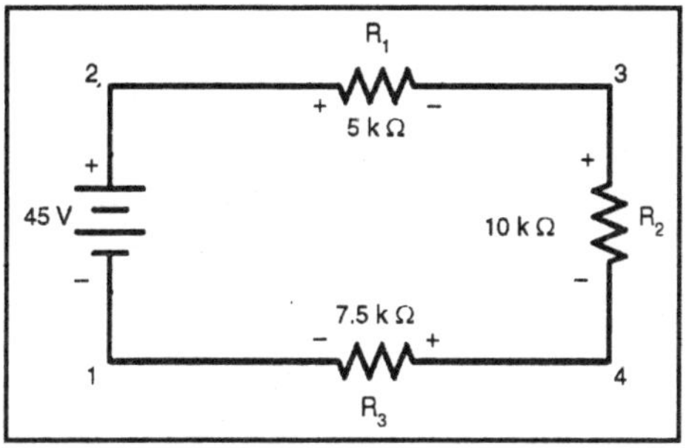

If we were to connect a voltmetre between points 2 and 1, red test lead to point 2 and black test lead to point 1, the metre would register +45 volts. Typically the"+"sign is not shown, but rather implied, for positive readings in digital metre displays. However, for this lesson the polarity of the voltage reading is very important and so I will show positive numbers explicitly:

$$E_{2-1} = +45\ V$$

When a voltage is specified with a double subscript (the characters"2–1" in the notation"E_{2-1}"), it means the voltage at the first point (2) as measured in reference to the second point (1). A voltage specified as"E_{cd}" would mean the voltage as indicated by a digital metre with the red test lead on point"c"and the black test lead on point"d": the voltage at"c" in reference to"d".

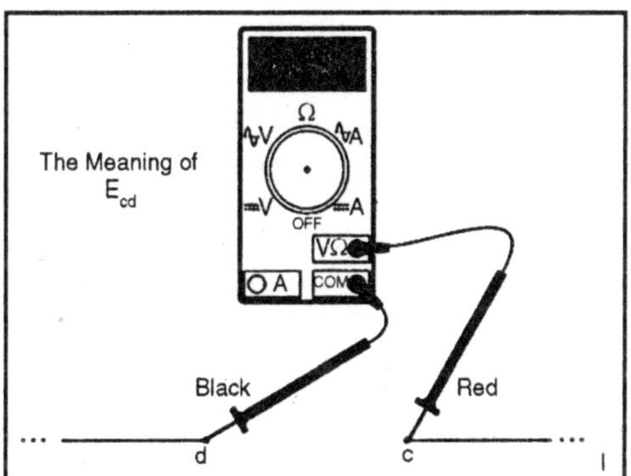

If we were to take that same voltmetre and measure the voltage drop across each resistor, stepping around the circuit in a clockwise direction with the red test lead of our metre on the point ahead and the black test lead on the point behind, we would obtain the following readings:

$E_{3-2} = -10\ V$
$E_{4-3} = -20\ V$
$E_{1-4} = -15\ V$

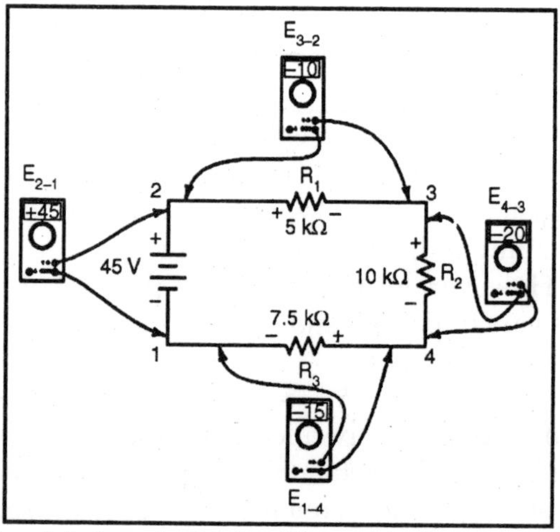

We should already be familiar with the general principle for series circuits stating that individual voltage drops add up to the total applied voltage but measuring voltage drops in this manner and paying attention to the polarity (mathematical sign) of the readings reveals another facet of this principle: that the voltages measured as such all add up to zero:

$$\begin{array}{rl} E_{2-1} & = +45\,V \\ E_{3-2} & = -10\,V \\ E_{4-3} & = -20\,V \\ +E_{1-4} & = -15\,V \\ \hline & 0V \end{array}$$

Voltage from point 2 to point 1
Voltage from point 3 to point 2
Voltage from point 4 to point 3
Voltage from point 1 to point 4

This principle is known as Kirchhoff's Voltage Law (discovered in 1847 by Gustav R. Kirchhoff, a German physicist), and it can be stated as such:

"The algebraic sum of all voltages in a loop must equal zero" By algebraic, I mean accounting for signs (polarities) as well as magnitudes. By loop, I mean any path traced from one point in a circuit around to other points in that circuit, and finally back to the initial point. In the above example the loop was formed by following points in this order: 1–2–3–4–1. It doesn't matter which point we start at or which direction we proceed in tracing the loop; the voltage sum will still equal zero. To demonstrate, we can tally up the voltages in loop 3–2–1–4–3 of the same circuit:

$$\begin{aligned} E_{2-3} &= +10\,V \\ E_{1-2} &= -45\,V \\ E_{4-1} &= +15\,V \\ +E_{3-4} &= +20\,V \\ \hline & 0\,V \end{aligned}$$

Voltage from point 2 to point 3
Voltage from point 1 to point 2
Voltage from point 4 to point 1
Voltage from point 3 to point 4

This may make more sense if we re-draw our example series circuit so that all components are represented in a straight line:

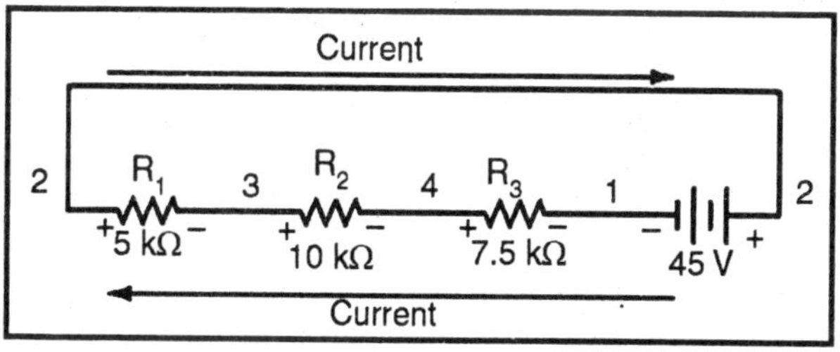

It's still the same series circuit, just with the components arranged in a different form. Notice the polarities of the resistor voltage drops with respect to the battery: the battery's voltage is negative on the left and positive on the right, whereas all the resistor voltage drops are oriented the other way: positive on the left and negative on the right. This is because the resistors are resisting the flow of electrons being pushed by the battery. In other words, the"push" exerted by the resistors against the flow of electrons must be in a direction opposite the source of electromotive force.

Here we see what a digital voltmetre would indicate across each component in this circuit, black lead on the left and red lead on the right, as laid out in horizontal fashion:

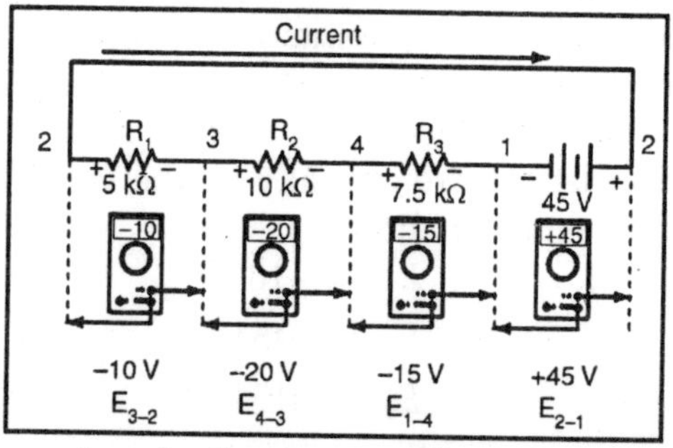

If we were to take that same voltmetre and read voltage across combinations of components, starting with only R_1 on the left and progressing across the whole string of components, we will see how the voltages add algebraically (to zero):

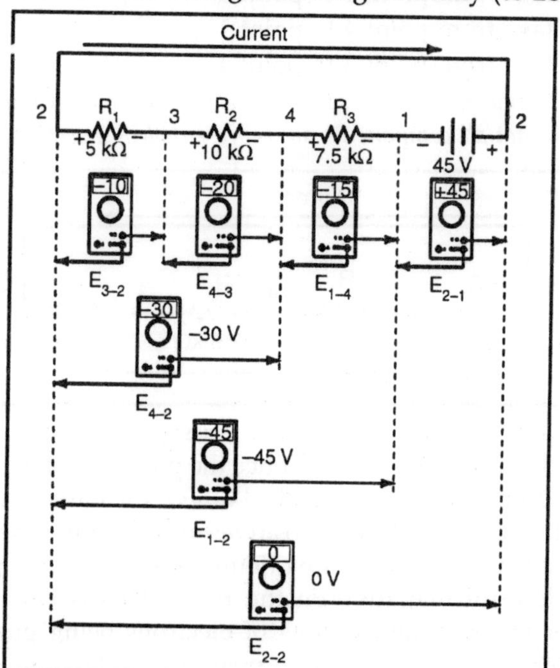

The fact that series voltages add up should be no mystery, but we notice that the polarity of these voltages makes a lot of difference in how the figures add. While reading voltage across R_1, R_1--R_2 and R_1--R_2--R_3 (I'm using a"double-dash" symbol"--" to represent the series connection between resistors R_1, R_2 and R_3), we see how the voltages measure successively larger (albeit negative) magnitudes, because the polarities of the individual voltage drops are in the same orientation (positive left, negative right).

The sum of the voltage drops across R_1, R_2 and R_3 equals 45 volts, which is the same as the battery's output, except that the battery's polarity is opposite that of the resistor voltage drops (negative left, positive right), so we end up with 0 volts measured across the whole string of components.

That we should end up with exactly 0 volts across the whole string should be no mystery, either. Looking at the circuit, we can see that the far left of the string (left side of R_1: point number 2) is directly connected to the far right of the string (right side of battery: point number 2), as necessary to complete the circuit. Since these two points are directly connected, they are electrically common to each other. And, as such, the voltage between those two electrically common points must be zero.

Kirchhoff's Voltage Law (sometimes denoted as KVL for short) will work for any circuit configuration at all, not just simple series.

Note how it works for this parallel circuit:

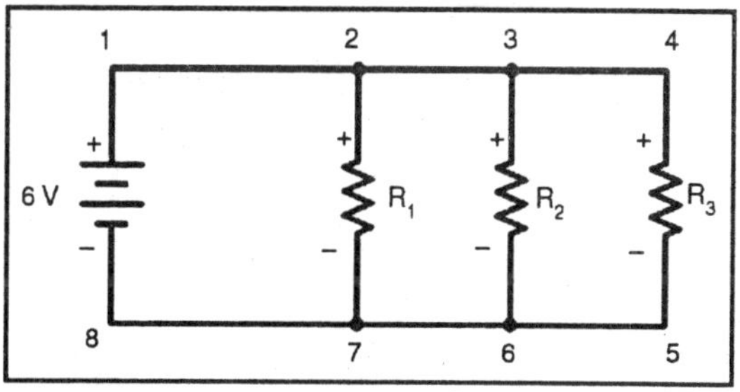

Being a parallel circuit, the voltage across every resistor is the same as the supply voltage: 6 volts. Tallying up voltages around loop 2–3–4–5–6–7–2. Note how I label the final (sum)

voltage as E_{2-2}. Since we began our loop-stepping sequence at point 2 and ended at point 2, the algebraic sum of those voltages will be the same as the voltage measured between the same point (E2–2), which of course must be zero.

$$E_{3-2} = 0\ V$$
$$E_{4-3} = 0\ V$$
$$E_{5-4} = -6\ V$$
$$E_{6-5} = 0\ V$$
$$E_{6-5} = 0\ V$$
$$E_{7-6} = 0\ V$$
$$+E_{2-7} = +6V$$
$$E_{2-2} = 0\ V$$

Voltage from point 3 to point 2
Voltage from point 4 to point 3
Voltage from point 5 to point 4
Voltage from point 6 to point 5
Voltage from point 7 to point 6
Voltage from point 2 to point 7

The fact that this circuit is parallel instead of series has nothing to do with the validity of Kirchhoff's Voltage Law. For that matter, the circuit could be a"black box"-its component configuration completely hidden from our view, with only a set of exposed terminals for us to measure voltage between -- and KVL would still hold true:

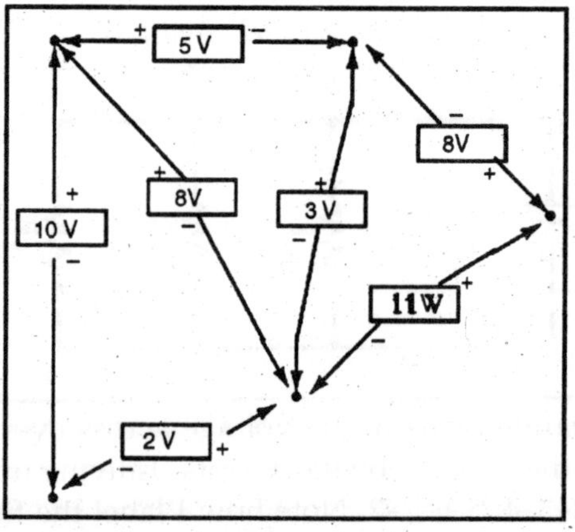

Try any order of steps from any terminal in the above diagram, stepping around back to the original terminal and you'll find that the algebraic sum of the voltages always equals zero.

Further more, the"loop" we trace for KVL doesn't even have to be a real current path in the closed-circuit sense of the word.

All we have to do to comply with KVL is to begin and end at the same point in the circuit, tallying voltage drops and polarities as we go between the next and the last point. Consider this absurd example, tracing"loop" 2–3–6–3–2 in the same parallel resistor circuit:

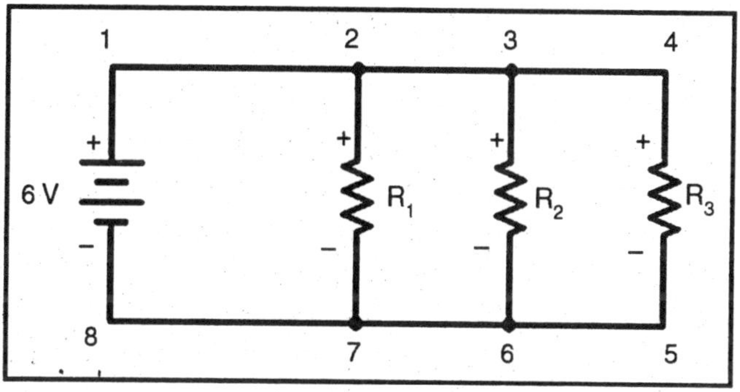

$$E_{3-2} = 0\ V$$
$$E_{6-3} = -6\ V$$
$$E_{3-6} = +6V$$
$$+E_{2-3} = 0\ V$$
$$E_{2-2} = 0\ V$$

Voltage from point 3 to point 2
Voltage from point 6 to point 3
Voltage from point 3 to point 6
Voltage from point 2 to point 3

KVL can be used to determine an unknown voltage in a complex circuit, where all other voltages around a particular"loop" are known. Take the following complex circuit (actually two series circuits joined by a single wire at the bottom) as an example:

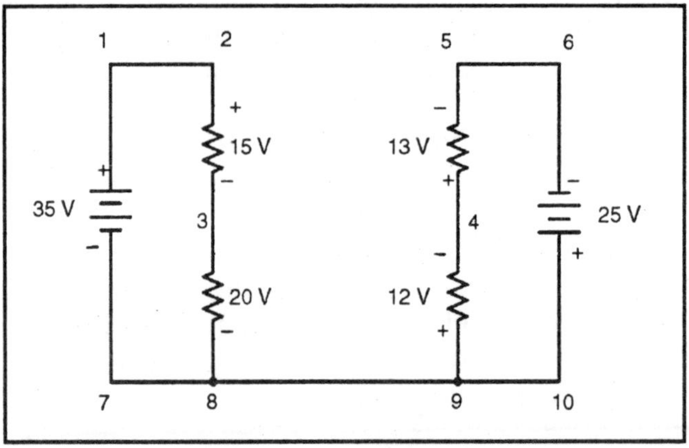

To make the problem simpler, I've omitted resistance values and simply given voltage drops across each resistor. The two series circuits share a common wire between them (wire 7–8–9–10), making voltage measurements between the two circuits possible. If we wanted to determine the voltage between points 4 and 3, we could set up a KVL equation with the voltage between those points as the unknown:

$$E_{4-3} + E_{9-4} + E_{8-9} + E_{3-8} = 0$$

$$E_{4-3} + 12 + 0 + 20 = 0$$

$$E_{4-3} + 32 = 0$$

$$E4 - 3 = -32V$$

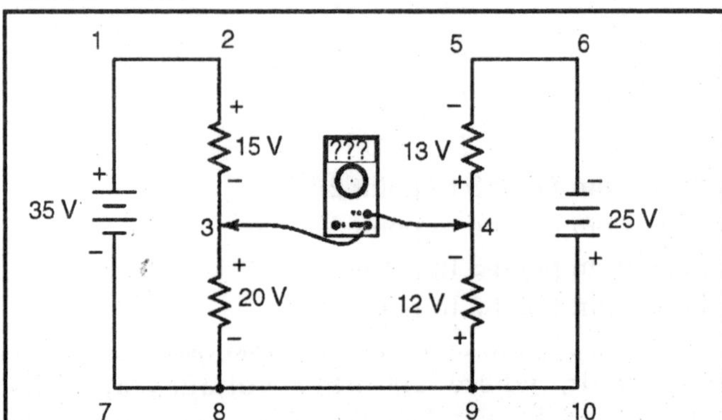

Measuring Voltage from Point 4 to Point 3 (Unknown Amount)

E_{4-3}

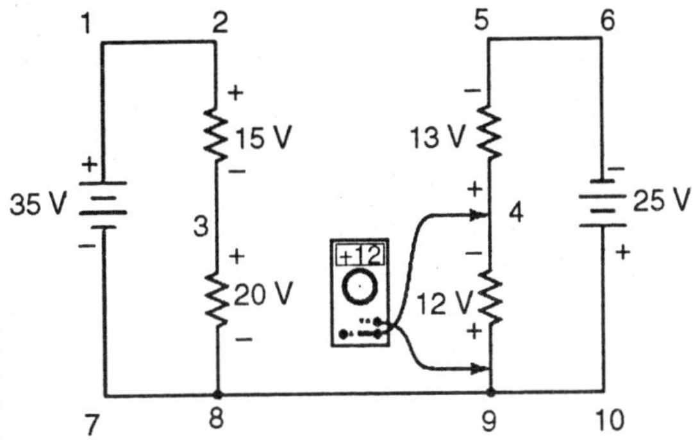

Measuring Voltage from Point 9 to Point 4 (+12 Volts)

$E_{4-3} + 12$

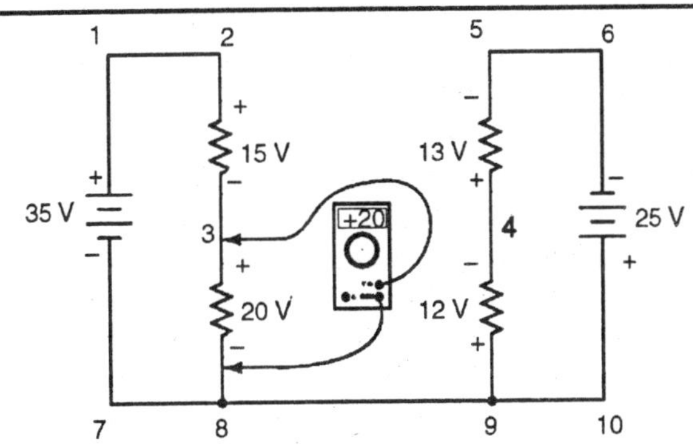

Measuring Voltage from Point 3 to Point 8 (+20Volts)

$E_{4-3} + 12 + 0 + 20 = 0$

Stepping around the loop 3–4–9–8–3, we write the voltage drop figures as a digital voltmetre would register them, measuring with the red test lead on the point ahead and black test lead on the point behind as we progress around the loop. Therefore, the voltage from point 9 to point 4 is a positive (+) 12 volts because the"red lead" is on point 9 and the"black lead" is on point 4.

The voltage from point 3 to point 8 is a positive (+) 20 volts because the"red lead" is on point 3 and the"black lead" is on point 8. The voltage from point 8 to point 9 is zero, of course because those two points are electrically common.

Our final answer for the voltage from point 4 to point 3 is a negative (–) 32 volts, telling us that point 3 is actually positive with respect to point 4, precisely what a digital voltmetre would indicate with the red lead on point 4 and the black lead on point 3:

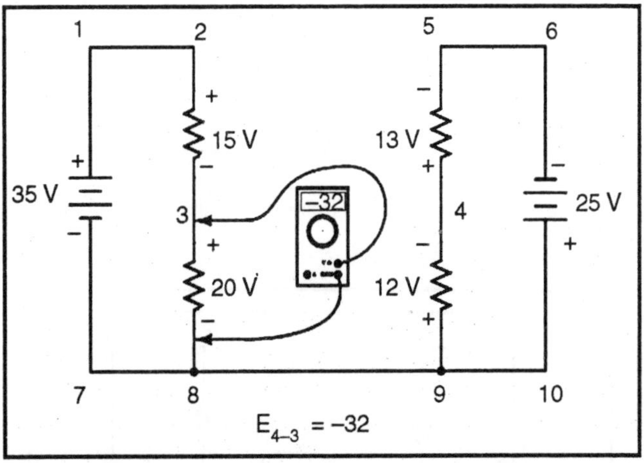

In other words, the initial placement of our"metre leads" in this KVL problem was"backwards". Had we generated our KVL equation starting with E3–4 instead of E4–3, stepping around the same loop with the opposite metre lead orientation, the final answer would have been E3–4 = +32 volts:

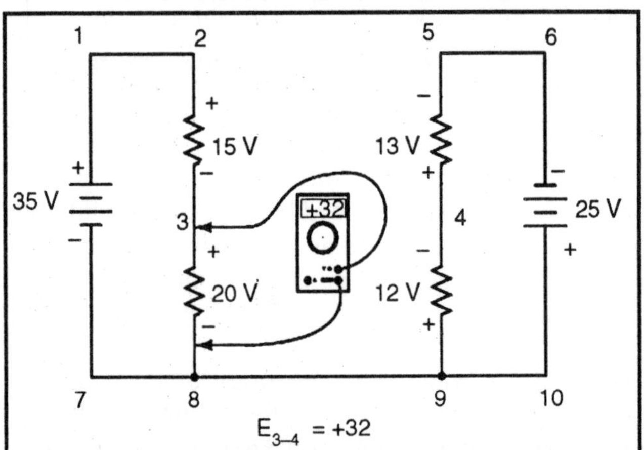

It is important to realise that neither approach is"wrong".

In both cases, we arrive at the correct assessment of voltage between the two points, 3 and 4: point 3 is positive with respect to point 4 and the voltage between them is 32 volts.

LOOP AND NODAL METHODS OF ANALYSIS

Having understood the fundamental laws of circuit theory (Ohm's law and Kirchhhoff's laws), we are now prepared to apply to develop two powerful techniques for circuit analysis: nodal analysis, which is based on a systematic application of Kirchhoff's current law (KCL) and mesh analysis which based on a systematic application of Kirchhoff's voltage law (KVL). The two techniques are so important that this chapter should be regarded as the most important in the book. Students are therefore encouraged to pay careful attention.

With the two techniques to be developed in this chapter we can analyse any linear circuit by obtaining a set of simultaneous equation that are then solved to obtain the required values of current or voltage. One method of solving simultaneous equations involves Cramer's rule, which allow us to calculate circuit variables as a quotient of determinants.

Finally, we apply the technique learned in this chapter to analyse transistor circuits.

NODAL ANALYSIS

Nodal analysis provides a general procedure circuits using node voltages as the circuit variables. Choosing node voltages instead of element voltages as circuit variables is convenient and reduces the number of equations one must solve simultaneously.

To simplify matters, we shall assume in this section that circuits do not contain voltage sources. Circuits that contain voltage sources will be analysed in the next section.

In nodal analysis, we are interested in finding the node voltages. Given a circuit with n nods without voltage sources, the nodal analysis of the circuit involves taking the following three steps.

Steps to determine node voltages:

- Select a node as the reference node. Assign voltages v1, v2, ..., v n–1 to remaining n–1 nodes. The voltages are referenced with respect to the reference node.
- Apply KCL to each of the n-1 nonreference nodes. Use Ohm's law to express the branch currents in term of node voltages.
- Solve the resulting simultaneous equations to obtain the unknown node voltages.

We shall now explain and apply these three steps.

The first step in nodal analysis is selecting a node as the reference or datum node. The reference node is commonly called the ground since it is assumed to have zero potential. A reference node is indicated by any of the three symbols in. The type of ground in is called a chassis ground and is used in devices where the case, enclosure, or chassis acts as reference point for all circuits. When the potential of the earth is used as reference, we use the earth ground in or. We shall always use the symbol in.

The Number of Nonreference nodes is Equal to the Number of Independent Equations that we will Derive.

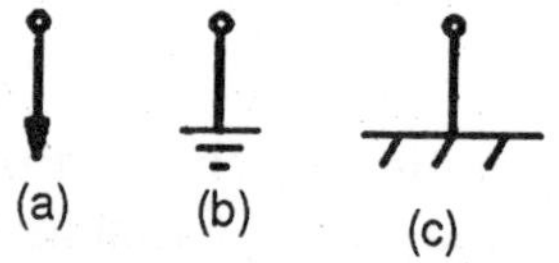

Fig. Common Symbols for Indicating a Reference node:

a) Common ground,
b) Ground,
c) Chassis ground.

Once we have selected a reference node, we assign voltage designations to nonreference nodes. Consider, for example, the circuit in Figure. Node 0 is the reference node (v = 0), while nodes 1 and 2 are assigned voltages v1 and v2, respectively. Keep in mind that the node voltages are defined with respect to the reference node. As illustrated in Figure each node voltage is the voltage rise from the reference node to the corresponding nonreference node or simply the voltage of that node with respect to the reference node.

As the second step, we apply KCL to each nonreference node in the circuit. To avoid putting too much information on the same circuit, the circuit is redrawn in, where we now add i1, i2, and i3 as the circuits through resistors R1, R2, and R3, respectively. At node 1, applying KCL gives,

I 1 = I 2 + i 1 + i 2

At node 2,

I 2 + i 2 = i 3

We now apply Ohm's to express the unknown currents i1, i2, and i3 in term of node voltages. The key idea to bear in mind is that, since resistance is a passive element, by the passive sign convention, current must always flow from a higher potential to a lower potential. Current flows from a higher potential to a lower potential in a resistor.

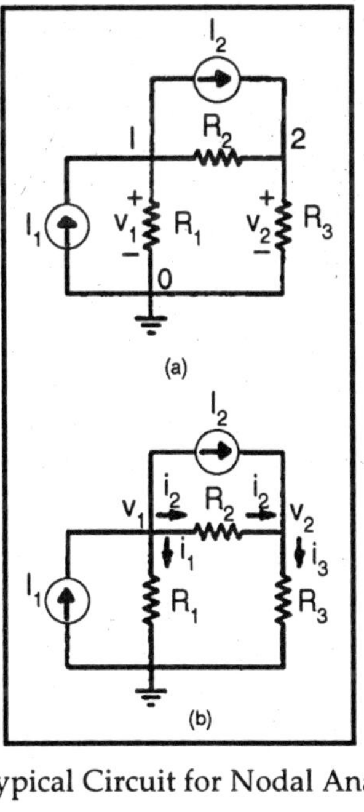

Fig. Typical Circuit for Nodal Analysis.

We can express this principle as,

i = v higher – v lower R

Note that this principle is in agreement with the way we define resistance in. With this in mind, we obtain from,

i1=v1–0R1 or i1=G1v1

i2=v1–v2R2

or

i2=G2(v1–v2)

i3=v2–0R3 or I3=G3v2

Substituting Equation 4 in Equation 1 and Equation 2 results, respectively, in

I 1 = I 2 + v 1 R 1 + v 1 – v 2 R 2

I 2 + v 1 – v 2 R 2 = v 2 R 3

In terms of the conductances, Equation 5 and Equation 6 become,

I 1 = I 2 + G 1 v 1 + G 2 (v 1 – v 2)

I 2 + G 2 (v 1 – v 2) = G 3 v 2

The third step in nodal analysis is to solve for the node voltages. If we apply KCL to n–1 nonreference nodes, we obtain n–1 simultaneous equations such as Equation and Equation or Equation and Equation. For the circuit of Figure, we solve Equation and Equation or Equation and Equation to obtain the node voltages v1 and v2 using any standard method, such as substitution method, elimination method, Cramer's rule or matrix inversion. To use either of the last two methods, one must cast the simultaneous equations in matrix form. For example, Equation and Equation can be cast in matrix form as

[G 1 + G 2 – G 2...– G 2 G 2 + G 3][v 1 v 2]=[I 1 – I 2... I 2]

which can be solved to get v1 and v2. (Reference) will be generalized. The simultaneous equations may also be solved using calculator or with software.

NODAL ANALYSIS WITH VOLTAGE SOURCES

We now consider how voltage sources effect nodal analysis. We use the circuit in below Figure for illustration. Consider the following two possibilities.

CASE 1: if a voltage source is connected between the reference node and a nonreference node, we supply set the voltage at the nonreference node equal to the voltage of the voltage source. In below Figure,

For example,

$$v1 = 10 \text{ V}$$

Thus our analysis is somewhat simplified by this knowledge of the voltage at this node.

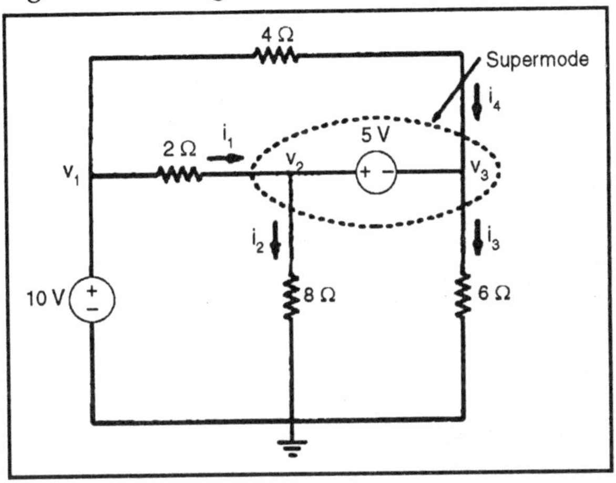

Fig. A Circuit with a Supernode.

If the voltage source (dependent or independent) is connected between two nonreference nodes, the two nonreference nodes form a generalized node or supernode; we apply both KCL and KVL to determine the node voltages.

A supernode is formed by enclosing a (dependent or independent) v1oltage source connected between two nonreference nodes and any elements connected in parallel with it.

In above Figure, nodes 2 and 3 form a supernode. We analyse a circuit with supernode using the same three steps mentioned in the previous section except that the supernodes are treated differently. Why? Because an essential component of nodal analysis is applying KCL, which requires knowing the current through each element. There is no way of knowing the current through a voltage source in advance. However, KCL must be satisfied at a supernode like any other node. Hence, at the supernode in above Figure,

$$i1 + i4 = i2 + i3$$

or

$$v1 - v22 + v1 - v34 = v2 - 08 + v3 - 06$$

To apply Kirchhoff's voltage law to the supernode in Figure, we redraw the circuit as shown in above Figure. Going around the loop in the clockwise direction gives,

$$-v2 + 5 + v3 = 0 - v2 - v3 = 5$$

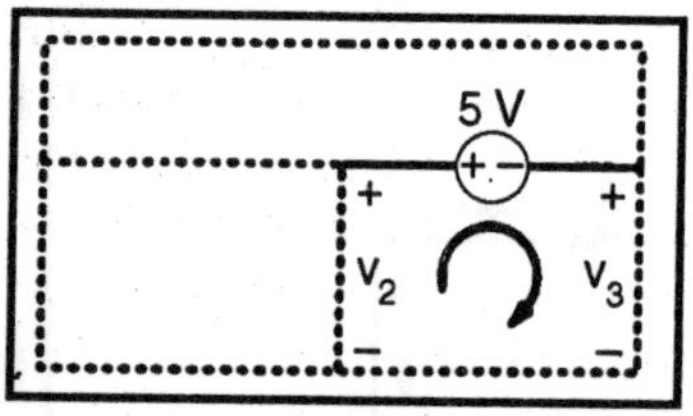

Fig. Applying KVL to a Supernode.

From Equation, Equation and Equation, we obtain the node voltages.

Note the following properties of a supernode:

- The voltage source inside the supernode provides a constraint equation needed to solve for the node voltage.
- A supernode has no voltage of it own.
- A supernode requires the application of both KCL and KVL.

MESH ANALYSIS

Mesh analysis is also known as loop analysis or the mesh current method. Mesh analysis provides another general procedure for analyzing circuits using mesh currents as the circuit variables. Using mesh currents instead of element currents as circuit variables is convenient and reduces the number of equations that must be solved simultaneously. Recall that a loop is closed path with no node passed more than once. A mesh is a loop that does not contain any other loop within it.

Nodal analysis applies KCL to find unknown voltages in a given circuit, while mesh analysis applies KVL to find unknown currents. Mesh analysis is not quite as general as nodal analysis because it is only applicable to a circuit that is planar. A planar circuit is one that can be drawn in a plane with no branches crossing one another; otherwise it is nonplanar. A circuit may have crossing branches and still be planar if it can be redrawn such that it has no crossing branches.

For example, the circuit in has two crossing branches, but it can be redrawn as in. Hence, the circuit in is planar. However, the circuit in Figure is nonplanar, because there is no way to redraw it and avoid the branches crossing. Nonplanar circuits can be handled using mesh analysis but they will not be considered in this text.

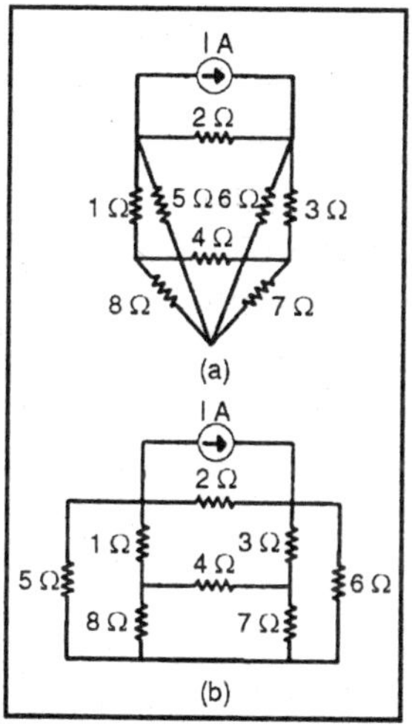

Fig. A Planar Circuit with Crossing Branches, b) the same Circuit Redrawn with no Crossing Branches.

To understand mesh analysis, we should first explain more about what we mean by a mesh. A mesh is a loop which does not contain any other loop within it.

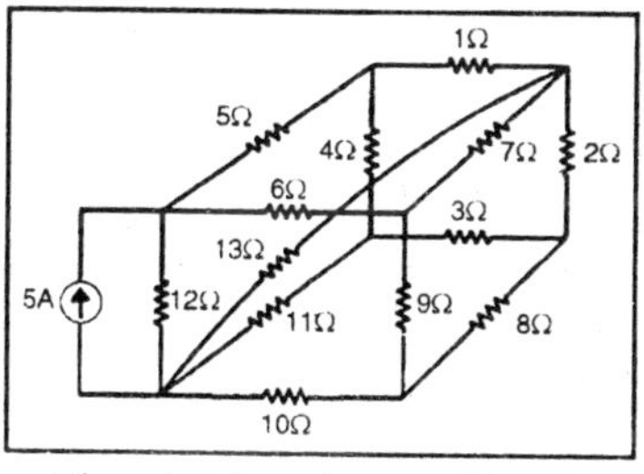

Fig. A Nonplanar Circuit

In below Figure, for example, paths abefa and bcdeb are meshes, but path abcdefa is not a mesh. The current through a mesh is known as mesh current. In mesh analysis, we are interested in applying KVL to find the mesh currents in a given circuit.

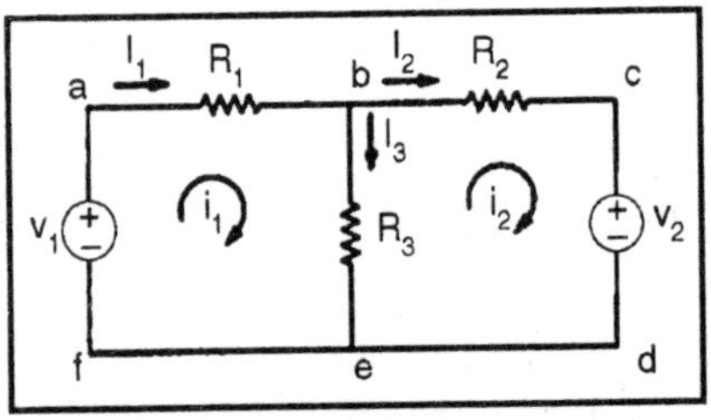

Fig. A Circuit with two Meshes.

In this section, we will apply mesh analysis to planar circuits that do not contain current sources. In the next section, we will consider circuits with current sources. In the mesh analysis of a circuit with *n* meshes, we take the following three steps.

Steps to determine mesh currents:

- Assign mesh currents i1, i2, ..., in to the n mesh.
- Apply KVL to each of n meshes. Use Ohm's law to express the voltages in term of the mesh currents.
- So the resulting n simultaneous equations to get the mesh currents.

To illustrate the steps, consider the circuit in Figure. The first step requires that mesh currents i1 and i2 are assigned to meshes 1 and 2. Although a mesh current may be assigned to each mesh in arbitrary direction, it is conventional to assume that each mesh current flows clockwise.

As the second step, we apply KVL to each mesh. Applying KVL to mesh 1, we obtain

$$-V1 + R1I1 + R3(i1 - i2) = 0$$

Or

$$(R1 + R3)i1 - R3i2 = V1$$

For mesh 2, applying KVL gives

$$R2i2 + V2 + R3(i1 - i2) = 0$$

or

$$-R3i1 + (R2 + R3)i2 = -V2$$

The third step is to solve for the mesh currents. Putting Equation 14 and Equation 15 in matrix form yields

[R 1 + R 2...? R 3 ? R 3 R 2 + R 3][i 1 i 2] = [V 1 -V 2]

Notice that the branch currents are different from the mesh current unless the mesh is isolated. To distinguish between the two types of currents, we use i for a mesh current and I for a branch current. The current elements I1, I2, and I3 are algebraic sums of the mesh currents. It is evident from above Figure that

I1 = i1

I2 = i2, I3 = i1–i2

MESH ANALYSIS WITH CURRENT SORCES

Applying mesh analysis to circuits containing current sources (dependent or independent) may appear complicated. But it is actually much easier than what we encountered in the previous section, because the presence of the current sources reduces the number of equations. Consider the following two possible cases.

When a current source exists only in one mesh: consider the circuit in above Figure, for example. We seti2=?5A and write a mesh equation for the other mesh in the usual way, that is,

– 10 + 4i 1 + 6 (i 1 – i 2) = 0, i 1 = – 2A

When a current source exists between two meshes: consider the circuit in, for example. We create a supermesh by excluding the current source and any elements connected in series with it. Thus, A supermesh results when two meshes have a (dependent or independent) current source in common.

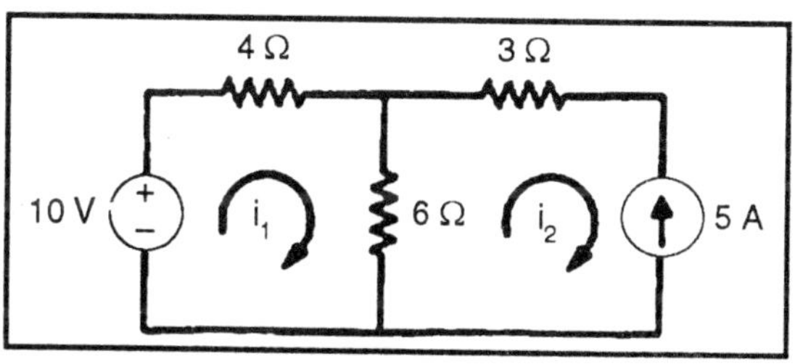

Fig. A Circuit with a Current Source.

As, w1e create a supermesh as the periphery of the two meshes and treat it differently. If a circuit has two or more supermeshes that intersect, they should be combined to form a larger supermesh.

Why treat the supermesh differently? Because mesh analysis applies KVL–which requires that we know the voltage across each branch-and we do not know the voltage across a current source in advance. However, a supermesh must satisfy KVL like any other mesh. Therefore, applying KVL to the supermesh gives

$$-20 + 6i\ 1 + 10\ i\ 2 + 4i\ 2 = 0$$

or

$$6i\ 1 + 14\ i\ 2 = 20$$

We apply KCL to a node in the branch where the two meshes intersect. Applying KCL to node 0 in gives

$$i\ 2 = i\ 1 + 6$$

Solving Equation and Equation, we get

$$i1 = -32A$$

and

$$I2 = 2.8A$$

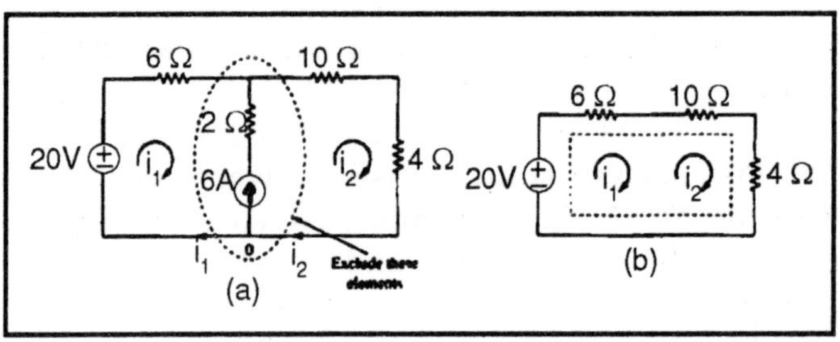

Fig. Two Meshes having a Current source in Common, b) a Supermesh Created by by Excluding the Current Source.

Note the following properties of a supermesh:

- The current source in the supermesh provides the constraint equation necessary to solve for the mesh currents.
- A supermesh has no current of its own.
- A supermesh requires the application of both KVL and KCL.

INSPECTION OF NODAL AND MESH ANALYSIS

This section presents a generalized procedure for nodal or mesh analysis. It is a shortcut approach based on mere inspection of a circuit.

When all sources in a circuit are independent current sources, we do not need to apply KCL to each node to obtain the node-voltage equations. We can obtain the equations by mere inspection of the circuit. As an example, let us reexamine, shown again in for convenience. The circuit has two nonreference nodes and the node equations were derived as

[G 1 + G 2... – G 2 ? G 2 G 2 + G 3][v 1 v 2] = [I 1– I 2... I 2]

Observe that each of the diagonal terms is sum of the conductances connected directly to node 1 or 2, while the off-diagonal terms are the negatives of the conductances connected between the nodes. Also, each term on the right-hand side of is the algebraic sum of the currents entering the node.

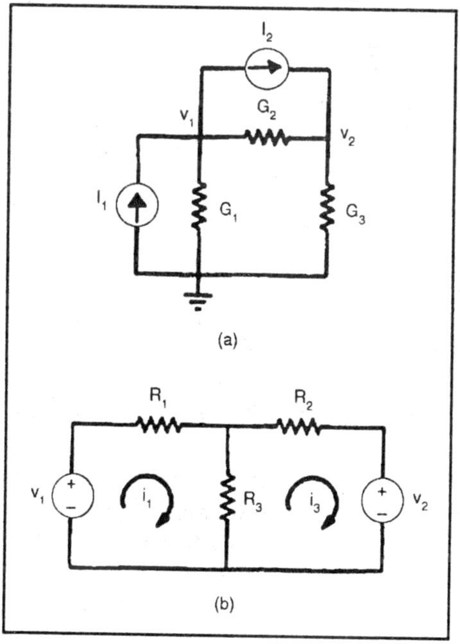

Fig. The circuit in the Circuit in.

In general, if a circuit with independent current sources has N nonreference nodes, the node-voltage equations can be written in term of the conductances as,

[G 11 G 12... G 1N – G N1 G N2... G NN][v 1 v N] = [i 1 I N... I 2]

or simply

$$Gv=i$$

where,

Gkk = Sum of the conductances connected to node k

Gki = Gik = negative of the sum of the conductances directly connecting nodes k and j, k≠j

vk = Unknown voltage at node k

ik = Sum of all independent current sources directly connected to node k, with currents entering the node treated as positive

G is called the conductance matrix: v is the output vector; and i is the input vector. (Reference) can be solved to obtain the unknown node voltages. Keep in mind that is valid for circuits only independent current sources and linear resistors.

Similarly, we can obtain mesh-current equations by inspection when a linear resistive circuit has only independent voltage sources. Consider the circuit in for convenience. The circuit has two non-reference nodes and the node equations were derived as,

[R 1 + R 2...– R 3 – R 3 R 2 + R 3][i 1 i 2]=[V 1–V 2]

We notice that each of the diagonal terms is the sum of the resistances in the related mesh, while each of the off-diagonal terms is the negative of the resistance common to meshes 1 and 2. Each term on the right-hand side of (Reference) is the algebraic sum taken clockwise of all independent voltage sources in the related mesh.

In general, if the circuit has N meshes, the mesh-current equations can be expressed in terms of the resistances as

[R 11 R 12... R 1N R N1 R N2 R NN][i 1 i N] = [v 1–v N]

or simply

Ri = v

where:

Rkk = Sum of the resistances in mesh k

Rki = Rik = negative of the sum of the resistances in common with meshes k and j, k ≠ j

ik = Unknown current for mesh k in the clockwise direction

vk = Sum taken clockwise of all independent voltage sources in mesh k, with voltage rise treated as positive

R is called the resistance matrix: v is the output vector; and i is the output vector; and v is the input vector. We can solve (Reference) to obtain the unknown mesh currents.

NODAL VERSUS MESH ANALYSIS

Both nodal and mesh analyses provide a systematic way of analyzing a complex network. Someone may ask: given a network to be analysed, how do we know which method is better or more efficient? The choice of the better method is dictated by two factors.

The first factor is the nature of the particular network. Networks that contain many series-connected elements, voltage sources, or supermeshes are more suitable for mesh analysis, whereas networks with parallel-connected elements, current sources or supernodes are more suitable for nodal analysis. Also, circuit with fewer nodes than meshes is better analysed using nodal analysis, while a circuit with fewer meshes than nodes is better analysed using mesh analysis. The key is to select the method that results in the smaller number of equations.

The second factor is the information required. If node voltages are required, it may be expedient to apply nodal analysis. If branch or mesh currents are required, it may be better to use mesh analysis.

It is helpful to be familiar with both methods of analysis, for at least two reasons. First, one method can be used to check the results from the other method, if possible. Second, since each method has its limitations, only one method may be suitable for a particular problem. For example, mesh analysis is the only method to use in analyzing transistor circuits.

But mesh analysis cannot easily be used to solve an op amp circuit, because there is no direct way to obtain the voltage across the op amp itself. For non planar networks, nodal analysis is the only option, because mesh analysis only applies to planar networks. Also, nodal analysis is more amenable to solution by computer, as it is easy to programme.

APPLICATIONS: DC TRANSISTOR CIRCUITS

Most of us deal with electronic products on a routine basis and have some experience with personal computers. A basic component for the integrated circuits found in these electronics and computers is the active, three terminal device known as the transistor. Understanding the transistor is essential before an engineer can start an electronic circuit design.

In Below Figure depicts various kinds of transistors commercially available. There are two basic types of transistors: bipolar junction transistors (BJTs) and field-effect transistor (FETs). Here, we consider only the BJTs, which were the first of the two and are still used today. Our objective is to present enough detail about the BJT to enable us to apply the techniques developed in this chapter to analyse dc transistor circuits.

There two types of BJTs: npn and pnp, with their circuit symbols as shown in. Each type has three terminals, designated as emitter (E), base (B) and collector (C). For the npn transistor, the currents and voltages of the transistor are specified as in. Applying KCL to gives

$$I E = I B + I C$$

where IE, IC, and IB are emitter, collector, and base currents, respectively.

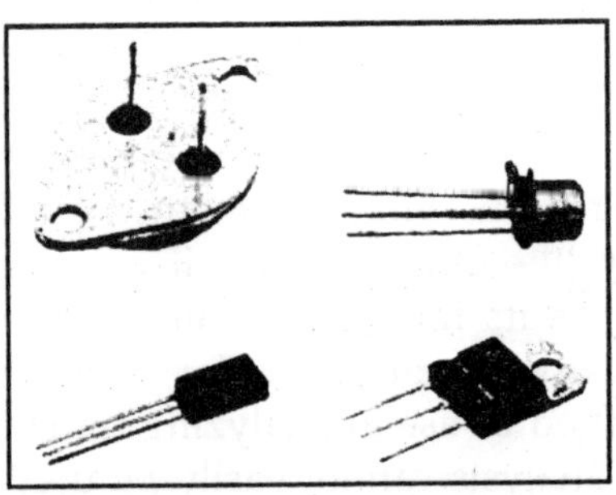

Fig. Various types of Transistor.

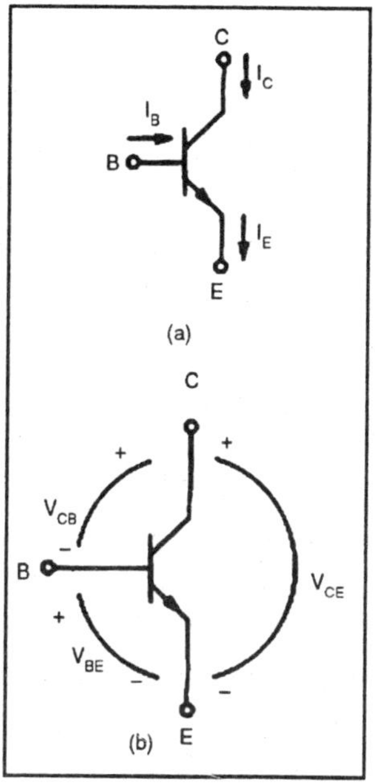

Fig. The terminal Variable of an npn Transistor: a) Currents, b) Voltages.

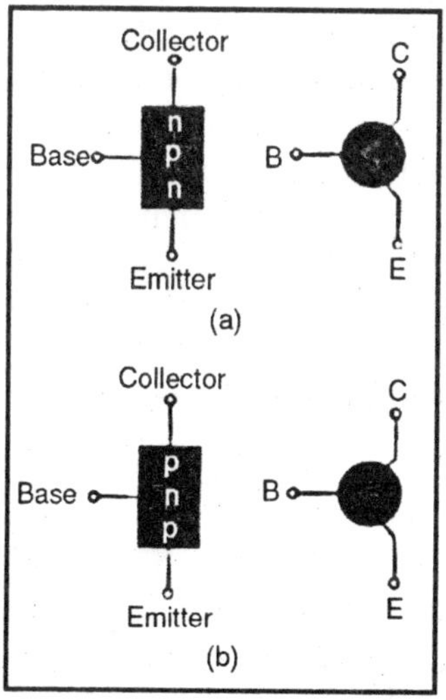

Fig. Two Types of BJTs and their Circuit Symbols: a) npn, b) pnp.

Similarly, applying KVL to gives,

V CE + V EB + V BC

Where VCE, VEB, and VBC are collector-emitter, emitter-base, and base-collector voltages. The BJT can operate in one of three modes: active, cutoff, and saturation. When transistors operate in the active mode, typically V ≈ 0.7V,

I C = aI E

Where α is called the common-base current gain. In Equation, α denotes the fraction of electrons injected by the emitter that are colleted by the collector. Also,

I C = bI B

Where β is known as the common-emitter current gain. The α and β are characteristic properties of a given transistor and assume constant values for that transistor. Typically, α takes values in the range of 0.98 to 0.999, while β takes values in the range 50 to 1000 from Equation to Equation, it is evident that,

I E = (1 + b) I B

and

b = a 1 a

These equations show that, in the active mode, the BJT can be modeled as a dependent current-controlled current source. Thus, in circuit analysis, the dc equivalent model in may be used to replace the npn transistor in. Since β in Equation is large, a small base current controls large currents in the output circuit. Consequently, the bipolar transistor can be serve as an amplifier, producing both current gain and voltage gain. Such amplifiers can be used to furnish a considerable amount of power to transducers such as loudspeakers or control motors.

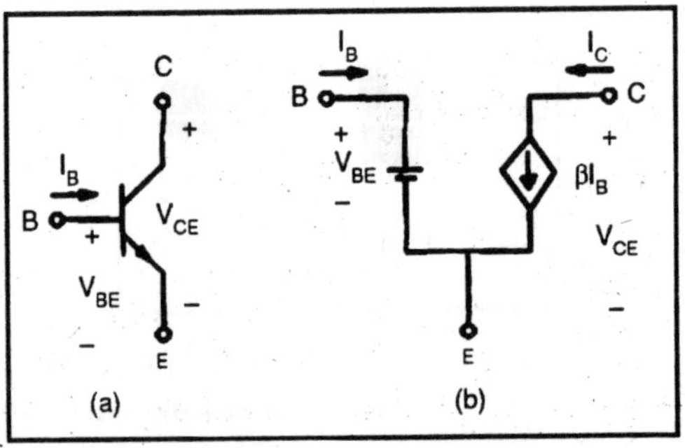

It should be observed in the following examples that one cannot directly analyse transistor circuits using nodal analysis because of the potential difference between the terminals of the transistor. Only when the transistor is replaced by its equivalent model, we can apply nodal analysis.

STAR-DELTA TRANSFORMATIONS

STAR AND DELTA TRANSFORMS

We can now solve simple series, parallel or bridge type resistive networks using Kirchoff's Circuit Laws, Mesh-current Analysis or Nodal-voltage Analysis techniques but in a balanced 3-phase circuit we can use different mathematical techniques to simplify the analysis of the circuit and thereby reduce the amount of math's involved which in itself is a good thing. Standard three phase circuits or networks take on two major forms with names that represent the way in which the resistances are connected, a Star connected Network which has the symbol of the letter, Y (wye) and a Delta connected Network which has the symbol of a triangle, Δ (delta). If a 3-phase, 3-wire supply or even a 3-phase load is connected in one type of configuration, it can be easily transformed or changed it into an equivalent configuration of the other type by using either the Star to Delta Transformation or Delta to Star Transformation process.

A resistive network consisting of three impedances can be connected together to form a T or"Tee" configuration but the network can also be redrawn to form a Star or Y type network as shown below.

T-connected and Star-connected Resistor Network.

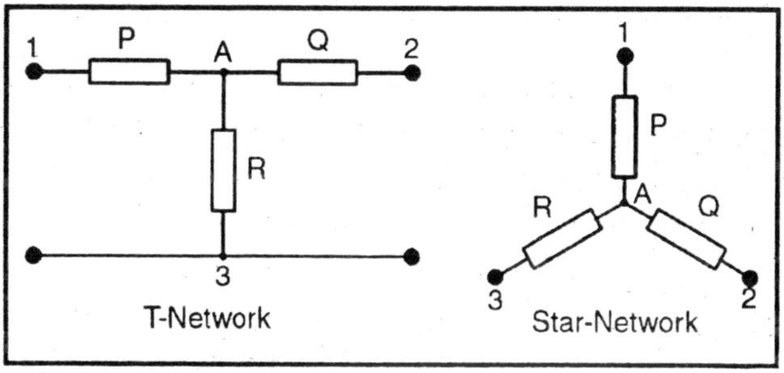

As we have already seen, we can redraw the T resistor network to produce an equivalent Star or Y type network. But we can also convert a Pi or π type resistor network into an equivalent Delta or Δ type network as shown below.

Pi-connected and Delta-connected Resistor Network.

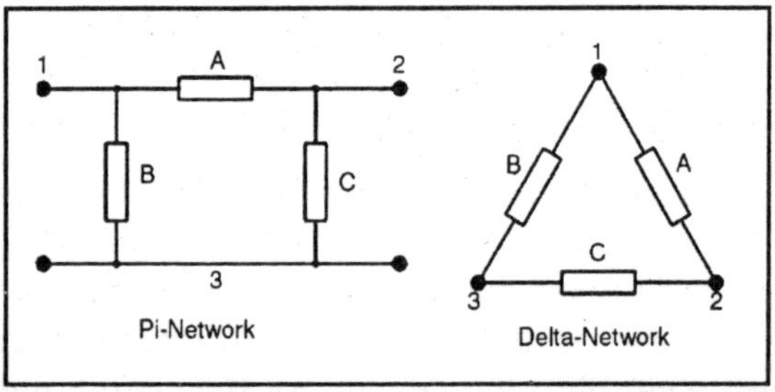

Having now defined exactly what is a Star and Delta connected network it is possible to transform the Y into an equivalent Δ network and also to convert a Δ into an equivalent Y network using aTransformation process. This process allows us to produce a mathematical relationship between the various resistors and their equivalents measured between the terminals 1–2, 1–3 or 2–3 for either a Star or Delta connected circuit. However, the resulting networks are only equivalent for voltages and currents external to the Star or Delta networks, as internally the voltages and currents are different but each network will consume the same amount of power and have the same power factor to each other.

Delta-Star Transformation: To convert a Delta network to an equivalent Star network we need to derive a transformation formula for equating the various resistors to each other between the various terminals. Consider the circuit below.

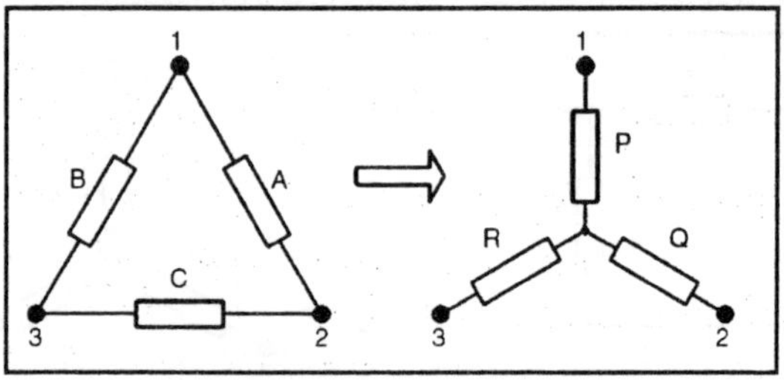

Fig. Delta to Star Network.

Compare the resistances between terminals 1 and 2.

P + Q = A in parallel with (B + C)

$$P+Q=\frac{A(B+C)}{A+B+C} \text{ ...EQ1}$$

Resistance between the terminals 2 and 3.

Q + R = C in parallel with (A + B)

$$Q+R=\frac{C(A+B)}{A+B+C} \text{ ...EQ2}$$

Resistance between the terminals 1 and 3.

P + R = B in parallel with (A + C)

$$P+R=\frac{B(A+C)}{A+B+C} \text{ ...EQ3}$$

This now gives us three equations and taking equation 3 from equation 2 gives:

$$EQ3-EQ2=(P+R)-(Q+R)$$

$$P+R=\frac{B(A+C)}{A+B+C}-Q+R=\frac{C(A+B)}{A+B+C}$$

$$\therefore P-Q=\frac{BA-CA}{A+B+C}-\frac{CA+CB}{A+B+C}$$

$$\therefore P-Q=\frac{BA-CA}{A+B+C}$$

Then, re-writing Equation 1 will give us:

$$P+Q=\frac{AB+AC}{A+B+C}$$

Adding together equation 1 and the result above of equation 3 minus equation 2 gives:

$$(P-Q)+(P+Q)$$

$$=\frac{BA-CA}{A+B+C}+\frac{AB+AC}{A+B+C}$$

$$=2P=\frac{2AB}{A+B+C}$$

From which gives us the final equation for resistor P as:

$$P=\frac{AB}{A+B+C}$$

Then to summarize a little the above maths, we can now say that resistor P in a Star network can be found as Equation 1 plus (Equation 3 minus Equation 2) or *Eq1 + (Eq3 - Eq2).*

Similarly, to find resistor Q in a Star network, is equation 2 plus the result of equation 1 minus equation 3 or Eq2 + (Eq1 - Eq3) and this gives us the transformation of Q as:

$$Q = \frac{AC}{A+B+C}$$

And again, to find resistor R in a Star network, is equation plus the result of equation minus equation 1 or *Eq3 + (Eq2 - Eq1)* and this gives us the transformation of *R* as:

$$Q = \frac{AC}{A+B+C}$$

When converting a Delta network into a Star network the denominators of all of the transformation formulas are the same: A + B + C, and which is the sum of all the Delta resistances. Then to convert any Delta connected network to an equivalent Star network we can summarized the above transformation equations as:

Delta to Star Transformations Equations

$$P = \frac{AB}{A+B+C}$$

$$Q = \frac{AC}{A+B+C}$$

$$R = \frac{BC}{A+B+C}$$

Example: Convert the following Delta Resistive Network into an equivalent Star Network.

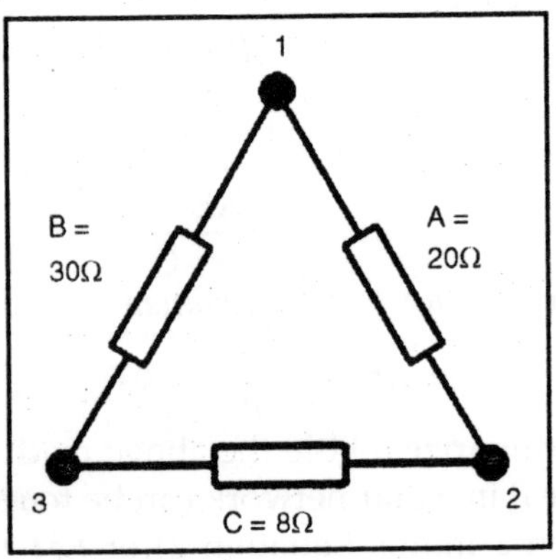

$$Q = \frac{AC}{A+B+C} = \frac{20 \times 80}{130} = 12.31\Omega$$

$$P = \frac{AB}{A+B+C} = \frac{20 \times 30}{130} = 4.61\Omega$$

$$R = \frac{BC}{A+B+C} = \frac{30 \times 80}{130} = 18.46\Omega$$

Star-Delta Transformation

We have seen above that when converting from a Delta network to an equivalent Star network that the resistor connected to one terminal is the product of the two Delta resistances connected to the same terminal, for example resistor P is the product of resistors A and B connected to terminal 1. By re-writing the previous formulas a little we can also find the transformation formulas for converting a resistive Star network to an equivalent Delta network as shown below.

Star to Delta Network.

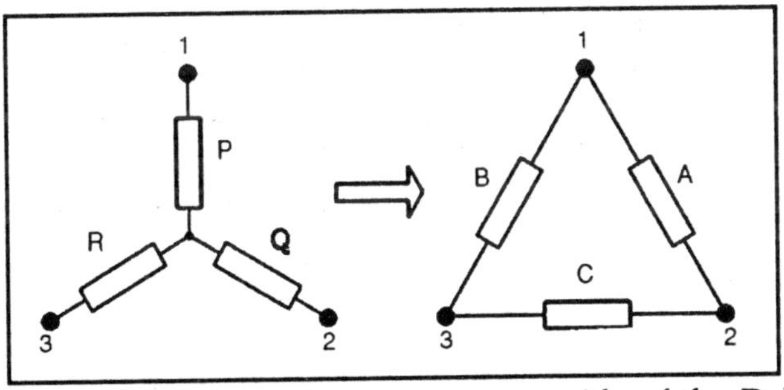

The value of the resistor on any one side of the Delta, Δ network is the sum of all the two-product combinations of resistors in the Star network divide by the Star resistor located"directly opposite"the Delta resistor being found. For example, resistor A is given as:

$$A = \frac{PQ + QR + RP}{R}$$

with respect to terminal 3 and resistor B is given as:

$$B = \frac{PQ + QR + RP}{Q}$$

with respect to terminal 2 with resistor C given as:

$$C = \frac{PQ + QR + RP}{P}$$

with respect to terminal 1.

By dividing out each equation by the value of the denominator we end up with three separate transformation formulas that can be used to convert any Delta resistive network into an equivalent Star network as given below.

Star to Delta Transformations Equations

$$A = \frac{PQ}{R} + Q + P$$

$$B = \frac{RP}{Q} + P + R$$

$$C = \frac{QR}{P} + Q + R$$

One final point about converting a Star resistive network to an equivalent Delta network. If all the resistors in the Star network are equal in value then the resultant resistors in the equivalent Delta network will be three times the value of the Star resistors and equal, giving: $R_{DELTA} = 3_{RSTAR}$

SUPERPOSITION THEOREM

DEFINITION

Superposition theorem is one of those strokes of genius that takes a complex subject and simplifies it in a way that makes perfect sense. A theorem like Millman's certainly works well, but it is not quite obvious why it works so well. Superposition, on the other hand, is obvious.

The strategy used in the Superposition1 Theorem is to eliminate all but one source of power within a network at a time, using series/parallel analysis to determine voltage drops (and/or currents) within the modified network for each power source separately. Then, once voltage drops and/or currents have been determined for each power source working separately, the values are all"superimposed"on top of each other (added algebraically) to find the actual voltage drops/

currents with all sources active. Let's look at our example circuit again and apply Superposition Theorem to it:

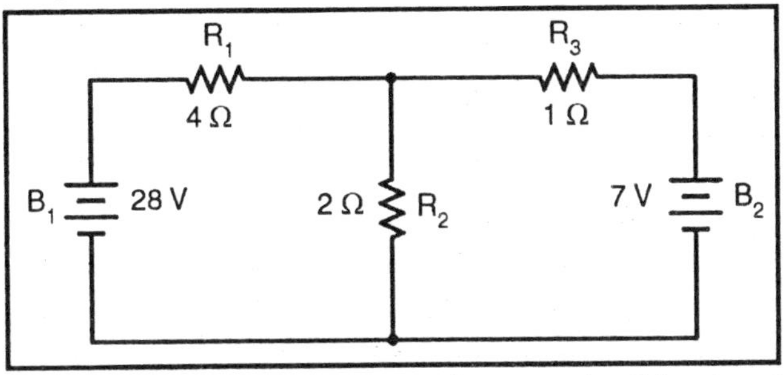

Since we have two sources of power in this circuit, we will have to calculate two sets of values for voltage drops and/or currents, one for the circuit with only the 28 volt battery in effect...

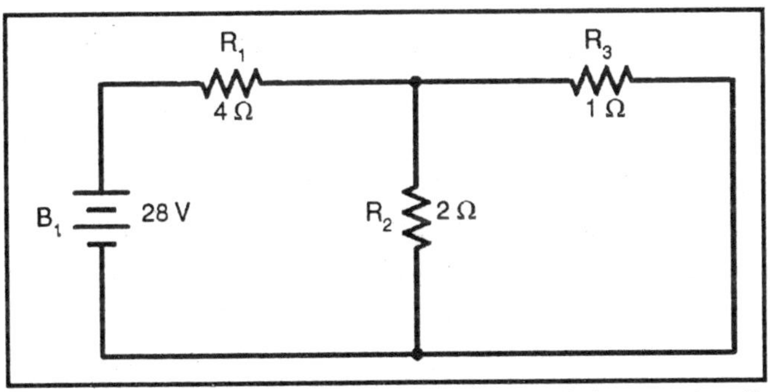

... and one for the circuit with only the 7 volt battery in effect:

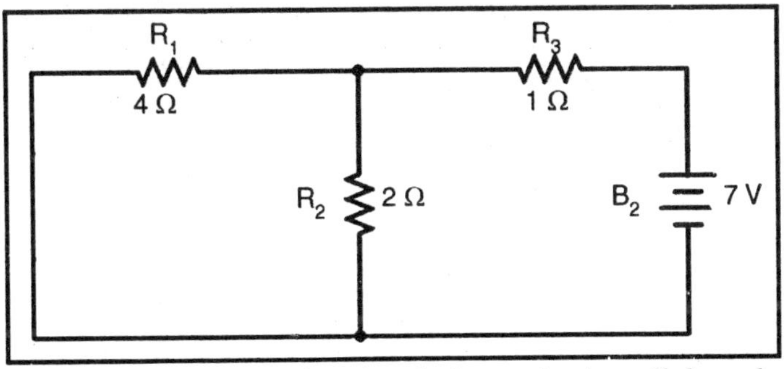

When re-drawing the circuit for series/parallel analysis with one source, all other voltage sources are replaced by wires (shorts) and all current sources with open circuits (breaks).

Since we only have voltage sources (batteries) in our example circuit, we will replace every inactive source during analysis with a wire.

Analyzing the circuit with only the 28 volt battery, we obtain the following values for voltage and current:

	R_1	R_2	R_3	$R_2//R_3$	$R_1 + R_2//R_3$ Total	
E	24	4	4	4	28	Volts
I	6	2	4	6	6	Amps
R	4	2	1	0.667	4.667	Ohms

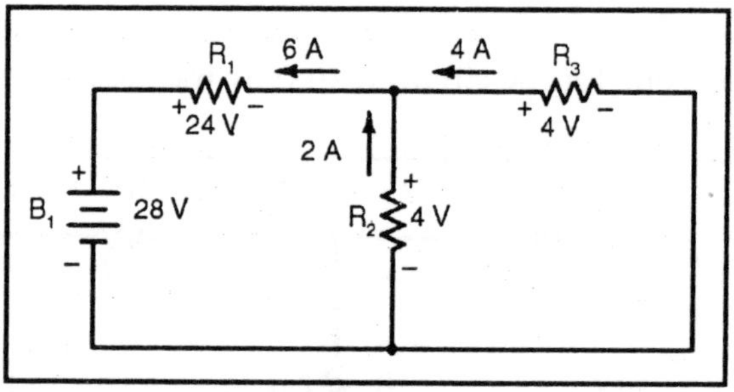

Analyzing the circuit with only the 7 volt battery, we obtain another set of values for voltage and current:

	R_1	R_2	R_3	$R_1//R_2$	$R_3 + R_1//R_2$ Total	
E	4	4	3	4	7	Volts
I	1	2	3	3	3	Amps
R	4	2	1	1.333	2.333	Ohms

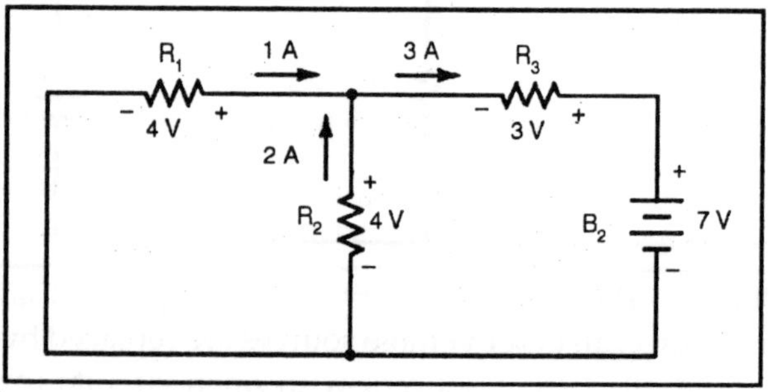

When superimposing these values of voltage and current, we have to be very careful to consider polarity (voltage drop) and direction (electron flow) as the values have to be added algebraically.

With 28 V Battery	With 7 V Battery	With Both Batteries
24 V + − E_{R1}	4 V − + E_{R1}	20 V + − E_{R1} 24 V−4 V =20 V
+ E_{R2} 4 V −	+ E_{R2} 4 V −	+ E_{R2} 8 V − 4 V + 4 V 8 V
4 V + − E_{R3}	3 V − + E_{R3}	1 V + − E_{R3} 4 V−3 V =1 V

Applying these superimposed voltage figures to the circuit, the end result looks something like this:

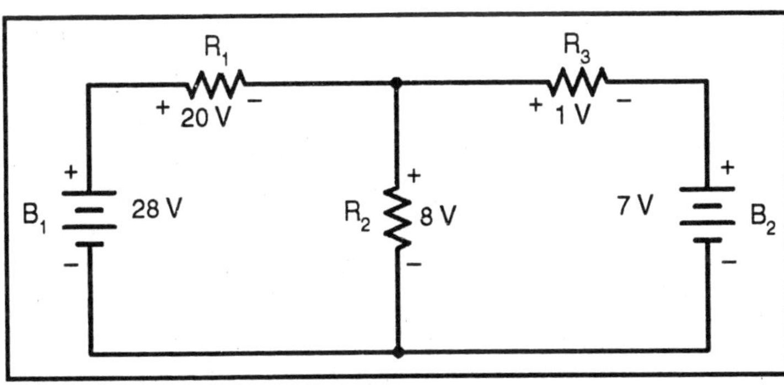

Currents add up algebraically as well and can either be superimposed as done with the resistor voltage drops, or simply calculated from the final voltage drops and respective resistances (I=E/R). Either way, the answers will be the same. Here I will show the superposition method applied to current:

With 28 V Battery	With 7 V Battery	With Both Batteries
I_{R1} ← 6 A	I_{R1} → 1 A	I_{R1} ← 5 A 6 A−4 A = 5 A
I_{R2} ↑ 2 A	I_{R2} ↑ 2 A	I_{R2} ↑ 4 A 2 A +2 A = 4 A
I_{R3} ← 4 A	I_{R3} → 3 A	I_{R3} ← 1 A 4 A−3 A=1 A

Once again applying these superimposed figures to our circuit:

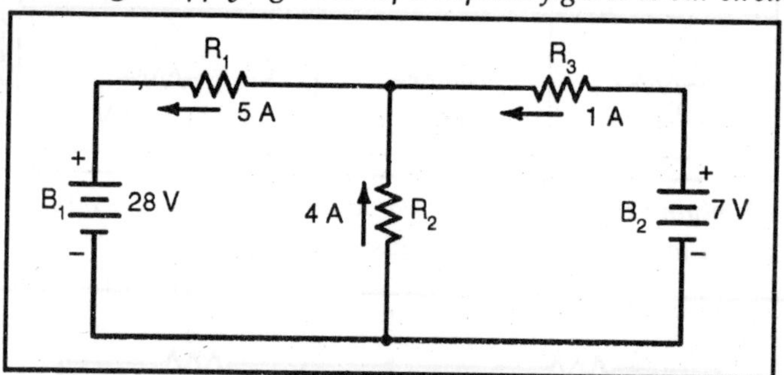

Quite simple and elegant, don't you think? It must be noted, though, that the Superposition Theorem works only for circuits that are reducible to series/parallel combinations for each of the power sources at a time (thus, this theorem is useless for analyzing an unbalanced bridge circuit) and it only works where the underlying equations are linear (no mathematical powers or roots). The requisite of linearity means that Superposition Theorem is only applicable for determining voltage and current, not power!!!

Power dissipations, being nonlinear functions, do not algebraically add to an accurate total when only one source is considered at a time. The need for linearity also means this Theorem cannot be applied in circuits where the resistance of a component changes with voltage or current. Hence, networks containing components like lamps (incandescent or gas-discharge) or varistors could not be analysed.

Another prerequisite for Superposition Theorem is that all components must be"bilateral," meaning that they behave the same with electrons flowing either direction through them. Resistors have no polarity-specific behaviour, and so the circuits we've been studying so far all meet this criterion.

The Superposition Theorem finds use in the study of alternating current (AC) circuits, and semiconductor (amplifier) circuits, where sometimes AC is often mixed (superimposed) with DC. Because AC voltage and current equations (Ohm's Law) are linear just like DC, we can use Superposition to analyse the circuit with just the DC power source, then just the AC power source, combining the results to tell what will happen with both AC and DC sources in effect. For now, though, Superposition will suffice as a break from having to do simultaneous equations to analyse a circuit.

THEVENIN'S THEOREM

DEFINITION

Thevenin's Theorem states that it is possible to simplify any linear circuit, no matter how complex to an equivalent circuit with just a single voltage source and series resistance connected to a load. The qualification of"linear" is identical to that found in the Superposition Theorem, where all the underlying equations must be linear (no exponents or roots). If we're dealing with passive components (such as resistors, and later, inductors and capacitors), this is true.

However, there are some components (especially certain gas-discharge and semiconductor components) which are nonlinear: that is their opposition to current changes with voltage and/or current. As such, we would call circuits containing these types of components, nonlinear circuits.

Thevenin's Theorem is especially useful in analyzing power systems and other circuits where one particular resistor in the circuit (called the"load" resistor) is subject to change and re-calculation of the circuit is necessary with each trial value of load resistance, to determine voltage across it and current through it. Let's take another look at our example circuit:

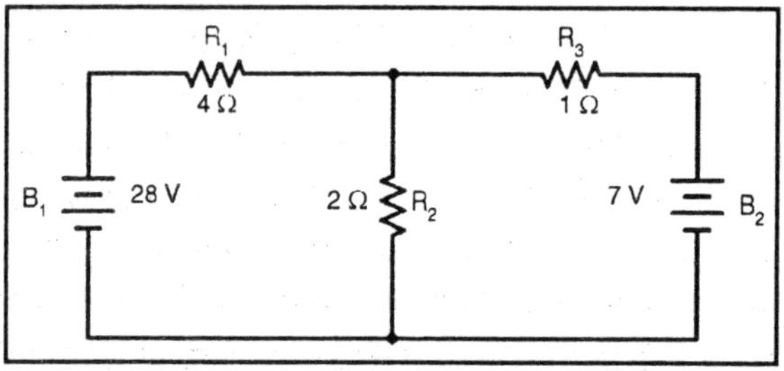

Let's suppose that we decide to designate R_2 as the"load" resistor in this circuit. We already have four methods of analysis at our disposal (Branch Current, Mesh Current, Millman's Theorem and Superposition Theorem) to use in determining voltage across R_2 and current through R_2 but each of these methods are time-consuming.

Imagine repeating any of these methods over and over again to find what would happen if the load resistance changed (changing load resistance is very common in power systems, as multiple loads get switched on and off as needed. the total resistance of their parallel connections changing depending on how many are connected at a time). This could potentially involve a lot of work! Thevenin's Theorem makes this easy by temporarily removing the load resistance from the original circuit and reducing what's left to an equivalent circuit composed of a single voltage source and series resistance. The load resistance can then be re-connected to this"Thevenin equivalent circuit" and calculations carried out as if the whole network were nothing but a simple series circuit:

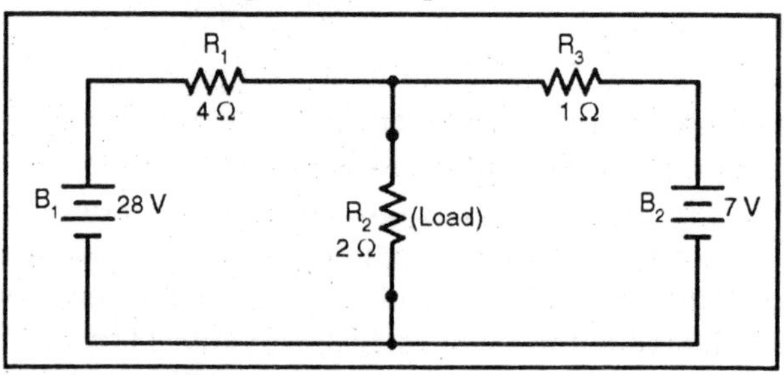

...after Thevenin conversion...

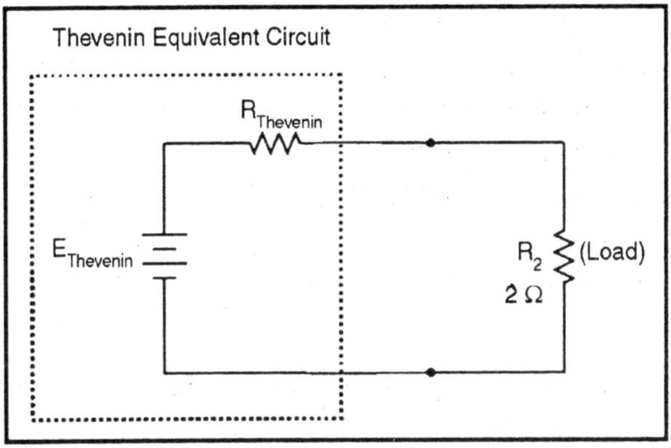

The"Thevenin Equivalent Circuit" is the electrical equivalent of B_1, R_1, R_3 and B_2 as seen from the two points where our load resistor (R_2) connects.

The Thevenin equivalent circuit, if correctly derived, will behave exactly the same as the original circuit formed by B_1, R_1, R_3 and B_2. In other words, the load resistor (R_2) voltage and current should be exactly the same for the same value of load resistance in the two circuits. The load resistor R_2 cannot"tell the difference" between the original network of B_1, R_1, R_3 and B_2 and the Thevenin equivalent circuit of $E_{Thevenin}$ and $R_{Thevenin}$, provided that the values for $E_{Thevenin}$ and $R_{Thevenin}$ have been calculated correctly.

The advantage in performing the"Thevenin conversion" to the simpler circuit, of course, is that it makes load voltage and load current so much easier to solve than in the original network. Calculating the equivalent Thevenin source voltage and series resistance is actually quite easy. First, the chosen load resistor is removed from the original circuit, replaced with a break (open circuit):

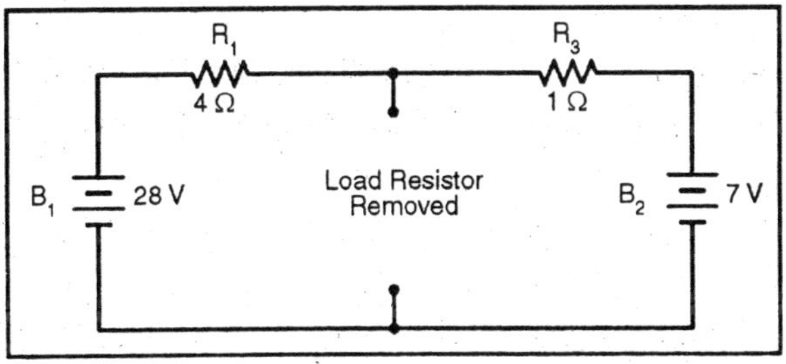

Next, the voltage between the two points where the load resistor used to be attached is determined. Use whatever analysis methods are at your disposal to do this. In this case, the original circuit with the load resistor removed is nothing more than a simple series circuit with opposing batteries, and so we can determine the voltage across the open load terminals by applying the rules of series circuits, Ohm's Law and Kirchhoff's Voltage Law:

	R_1	R_3	Total	
E	16.8	4.2	21	Volts
I	4.2	4.2	4.2	Amps
R	4	1	5	Ohms

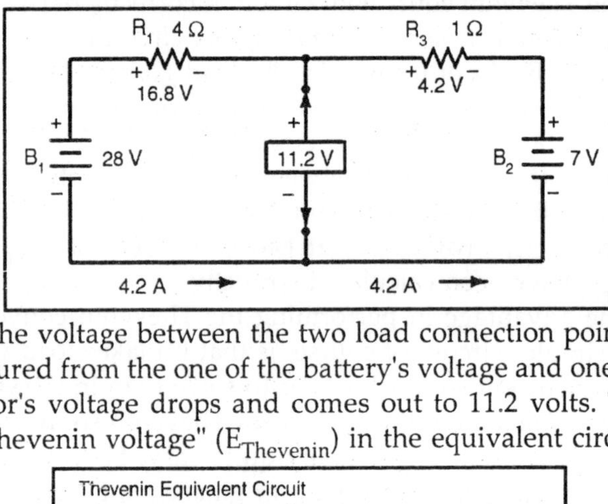

The voltage between the two load connection points can be figured from the one of the battery's voltage and one of the resistor's voltage drops and comes out to 11.2 volts. This is our"Thevenin voltage" ($E_{Thevenin}$) in the equivalent circuit:

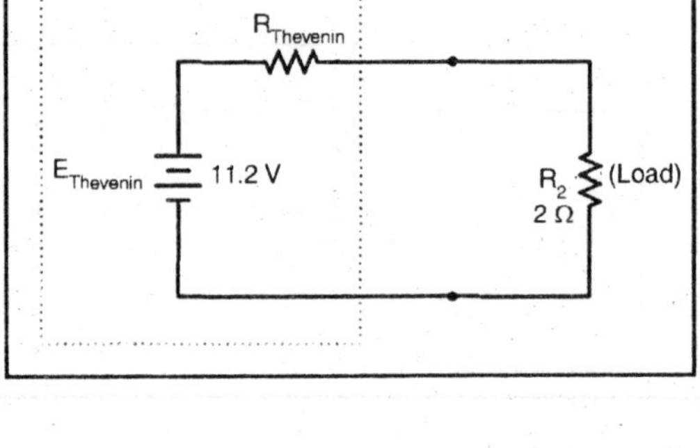

To find the Thevenin series resistance for our equivalent circuit, we need to take the original circuit (with the load resistor still removed), remove the power sources (in the same style as we did with the Superposition Theorem: voltage sources replaced with wires and current sources replaced with breaks) and figure the resistance from one load terminal to the other:

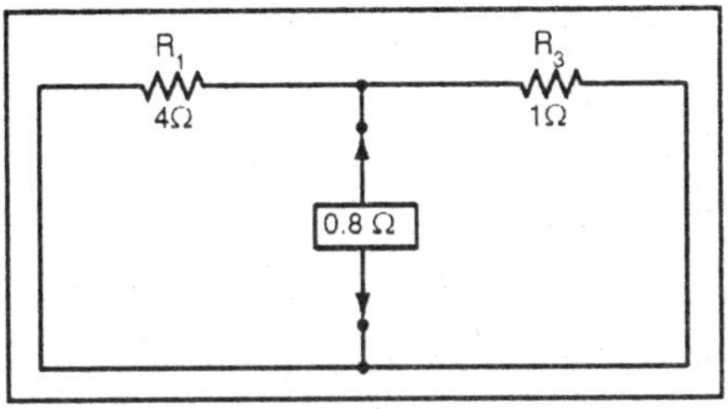

With the removal of the two batteries, the total resistance measured at this location is equal to R_1 and R_3 in parallel: 0.8 Ω. This is our"Thevenin resistance" ($R_{Thevenin}$) for the equivalent circuit:

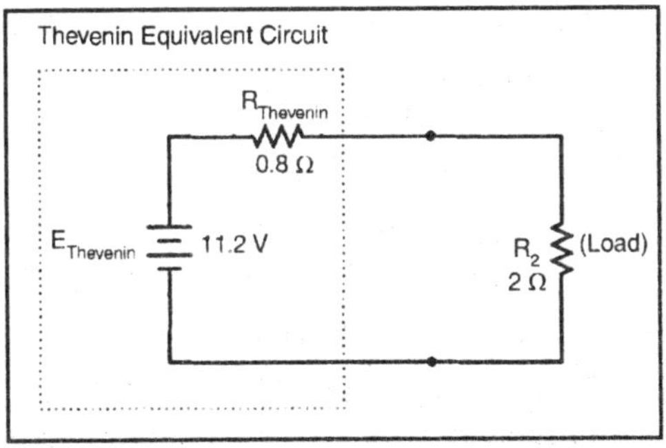

With the load resistor (2Ω) attached between the connection points, we can determine voltage across it and current through it as though the whole network were nothing more than a simple series circuit:

	$R_{Thevenin}$	R_{Load}	Total	
E	3.2	8	11.2	Volts
I	4	4	4	Amps
R	0.8	2	2.8	Ohms

Notice that the voltage and current figures for R_2 (8 volts, 4 amps) are identical to those found using other methods of analysis.

Also notice that the voltage and current figures for the Thevenin series resistance and the Thevenin source (total) do not apply to any component in the original, complex circuit. Thevenin's Theorem is only useful for determining what happens to a single resistor in a network: the load.

The advantage, of course, is that you can quickly determine what would happen to that single resistor if it were of a value other than 2Ω without having to go through a lot of analysis again. Just plug in that other value for the load resistor into the Thevenin equivalent circuit and a little bit of series circuit calculation will give you the result.

NORTON'S THEOREM

DEFINITION

Norton's Theorem states that it is possible to simplify any linear circuit, no matter how complex, to an equivalent circuit with just a single current source and parallel resistance connected to a load. Just as with Thevenin's Theorem, the qualification of"linear" is identical to that found in the Superposition Theorem: all underlying equations must be linear (no exponents or roots).

Contrasting our original example circuit against the Norton equivalent: it looks something like this:

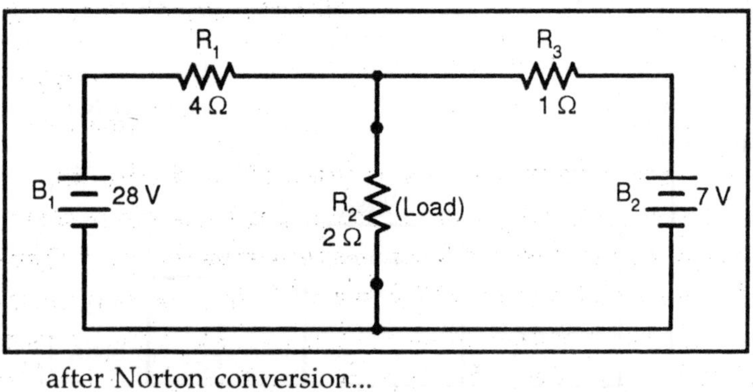

after Norton conversion...

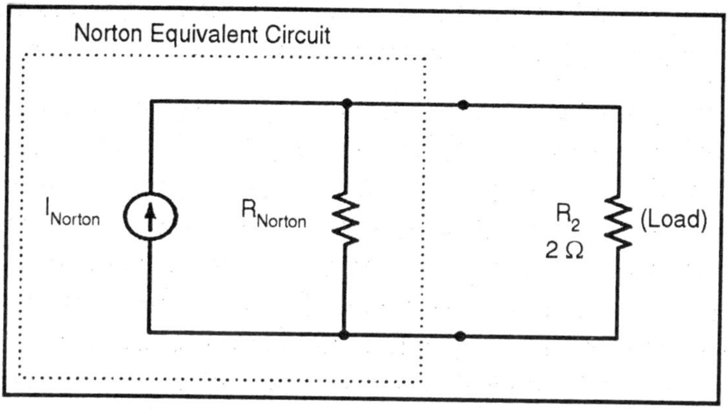

Remember that a current source is a component whose job is to provide a constant amount of current, outputting as much or as little voltage necessary to maintain that constant current.

As with Thevenin's Theorem, everything in the original circuit except the load resistance has been reduced to an equivalent circuit that is simpler to analyse. Also similar to Thevenin's Theorem are the steps used in Norton's Theorem to calculate the Norton source current (I_{Norton}) and Norton resistance (R_{Norton}).

As before, the first step is to identify the load resistance and remove it from the original circuit:

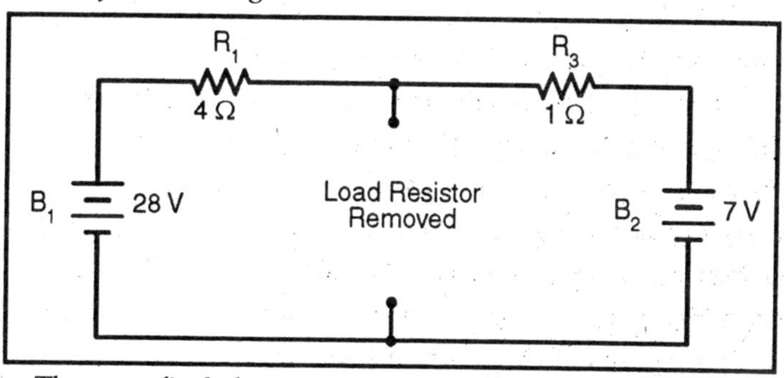

Then, to find the Norton current (for the current source in the Norton equivalent circuit), place a direct wire (short) connection between the load points and determine the resultant current. Note that this step is exactly opposite the respective step in Thevenin's Theorem, where we replaced the load resistor with a break (open circuit):

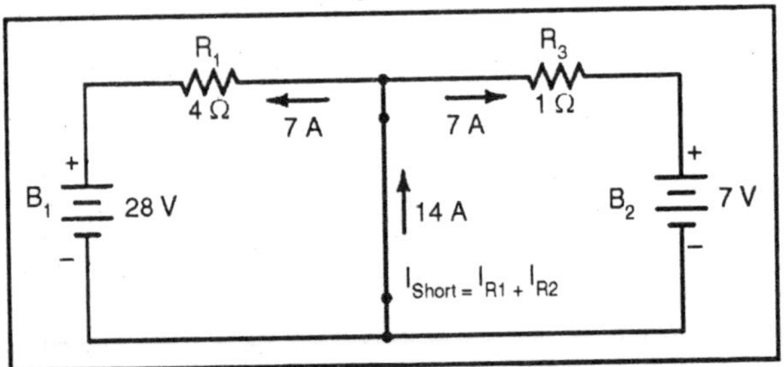

With zero voltage dropped between the load resistor connection points, the current through R_1 is strictly a function of B_1's voltage and R_1's resistance: 7 amps (I=E/R). Likewise, the current through R_3 is now strictly a function of B_2's voltage and R_3's resistance: 7 amps (I=E/R). The total current through the short between the load connection points is the sum of these two currents: 7 amps + 7 amps = 14 amps. This figure of 14 amps becomes the Norton source current (I_{Norton}) in our equivalent circuit:

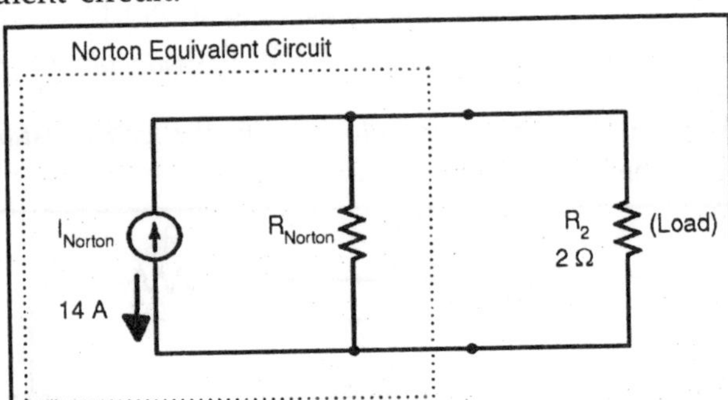

Remember, the arrow notation for a current source points in the direction opposite that of electron flow. Again, apologies for the confusion. For better or for worse, this is standard electronic symbol notation. Blame Mr. Franklin again!

To calculate the Norton resistance (R_{Norton}), we do the exact same thing as we did for calculating Thevenin resistance ($R_{Thevenin}$): take the original circuit (with the load resistor still removed), remove the power sources (in the same style as we did with the Superposition Theorem: voltage sources replaced with wires and current sources replaced with breaks) and figure total resistance from one load connection point to the other:

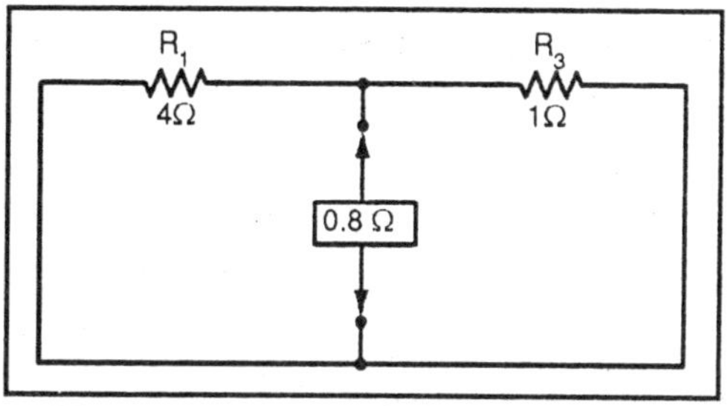

Now our Norton equivalent circuit looks like this:

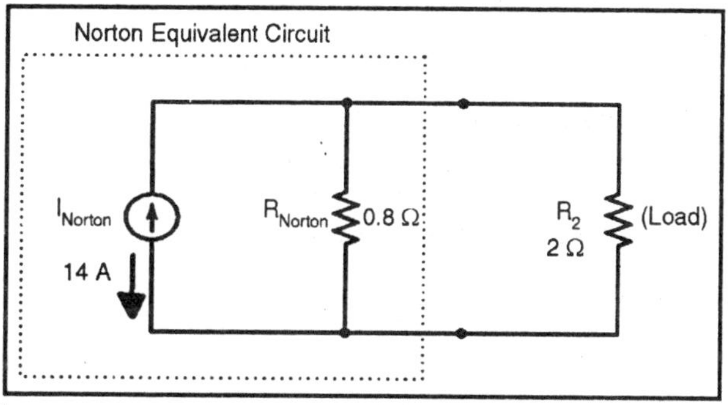

If we re-connect our original load resistance of 2Ω, we can analyse the Norton circuit as a simple parallel arrangement:

	R_{Norton}	R_{Load}	Total	
E	8	8	8	Volts
I	10	4	14	Amps
R	0.8	2	571.43m	Ohms

As with the Thevenin equivalent circuit, the only useful information from this analysis is the voltage and current values for R_2; the rest of the information is irrelevant to the original circuit. However, the same advantages seen with Thevenin's Theorem apply to Norton's as well: if we wish to analyse load resistor voltage and current over several different values of load resistance, we can use the Norton equivalent circuit again and again, applying nothing more complex than simple parallel circuit analysis to determine what's happening with each trial load.

MAXIMUM POWER TRANSFER THEOREM

DEFINITION

The Maximum Power Transfer Theorem is not so much a means of analysis as it is an aid to system design. Simply stated, the maximum amount of power will be dissipated by a load resistance when that load resistance is equal to the Thevenin/Norton resistance of the network supplying the power. If the load resistance is lower or higher than the Thevenin/Norton resistance of the source network, its dissipated power will be less than maximum.

This is essentially what is aimed for in stereo system design, where speaker"impedance" is matched to amplifier"impedance" for maximum sound power output. Impedance, the overall opposition to AC and DC current, is very similar to resistance and must be equal between source and load for the greatest amount of power to be transferred to the load. A load impedance that is too high will result in low power output. A load impedance that is too low will not only result in low power output, but possibly overheating of the amplifier due to the power dissipated in its internal (Thevenin or Norton) impedance.

Taking our Thevenin equivalent example circuit, the Maximum Power Transfer Theorem tells us that the load resistance resulting in greatest power dissipation is equal in value to the Thevenin resistance (in this case, 0.8 Ω): *With this value of load resistance, the dissipated power will be 39.2 watts:*

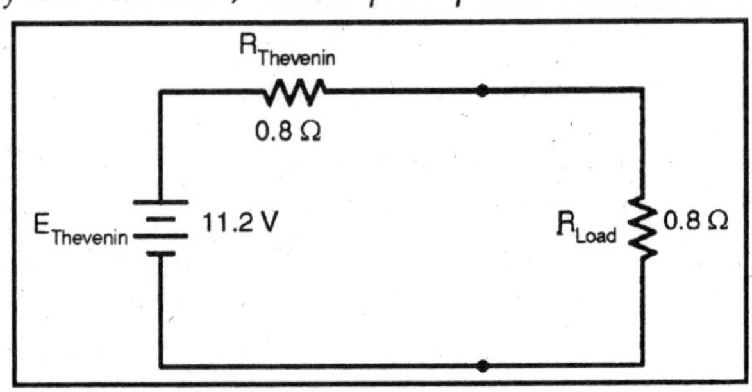

If we were to try a lower value for the load resistance (0.5 Ω instead of 0.8 Ω, for example), our power dissipated by the load resistance would decrease:

	$R_{Thevenin}$	R_{Load}	Total	
E	5.6	5.6	11.2	Volts
I	7	7	7	Amps
R	0.8	0.8	1.6	Ohms
P	39.2	39.2	78.4	Watts

Power dissipation increased for both the Thevenin resistance and the total circuit but it decreased for the load resistor. Likewise, if we increase the load resistance (1.1 Ω instead of 0.8 Ω, for example), power dissipation will also be less than it was at 0.8 Ω exactly:

	$R_{Thevenin}$	R_{Load}	Total	
E	6.892	4.308	11.2	Volts
I	8.615	8.615	8.615	Amps
R	0.8	0.5	1.3	Ohms
P	59.38	37.11	96.49	Watts

If you were designing a circuit for maximum power dissipation at the load resistance, this theorem would be very useful. Having reduced a network down to a Thevenin voltage and resistance (or Norton current and resistance), you simply set the load resistance equal to that Thevenin or Norton equivalent (or vice versa) to ensure maximum power dissipation at the load. Practical applications of this might include stereo amplifier design (seeking to maximize power delivered to speakers) or electric vehicle design (seeking to maximize power delivered to drive motor).

Chapter 3

Analysis of Single Phase AC Circuits

AC FUNDAMENTALS

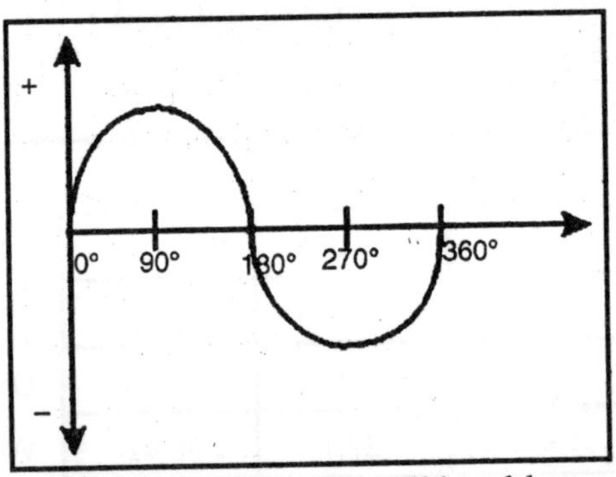

Objective: From memory, you will be able to explain and/or illustrate the generation of alternating current (AC) and voltage. AC is the generation used in most electrical utilities. A good understanding of these principles will assist in future line work.

ALTERNATING CURRENT

Alternating Current vs. Direct Current

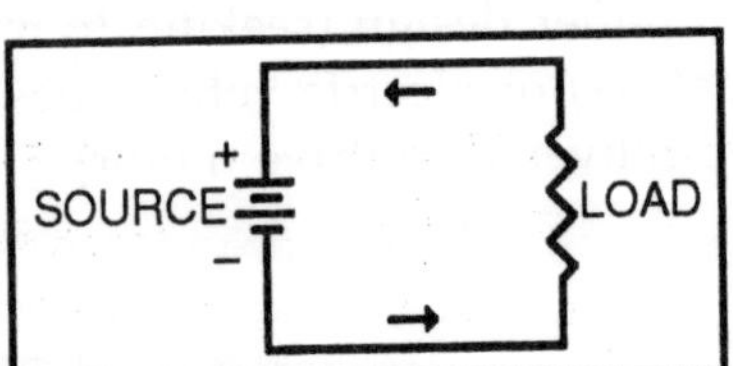

The figure shows the schematic diagram of a very basic DC circuit. It consists of nothing more than a source (a producer of electrical energy) and a load (whatever is to be powered by that electrical energy).

The source can be any electrical source: a chemical battery, an electronic power supply, a mechanical generator or any other possible continuous source of electrical energy. For simplicity, we represent the source in this figure as a battery.

At the same time, the load can be any electrical load: A light bulb, electronic clock or watch, electronic instrument, or anything else that must be driven by a continuous source of electricity. The figure here represents the load as a simple resistor.

Regardless of the specific source and load in this circuit, electrons leave the negative terminal of the source, travel through the circuit in the direction shown by the arrows, and eventually return to the positive terminal of the source. This action continues for as long as a complete electrical circuit exists.

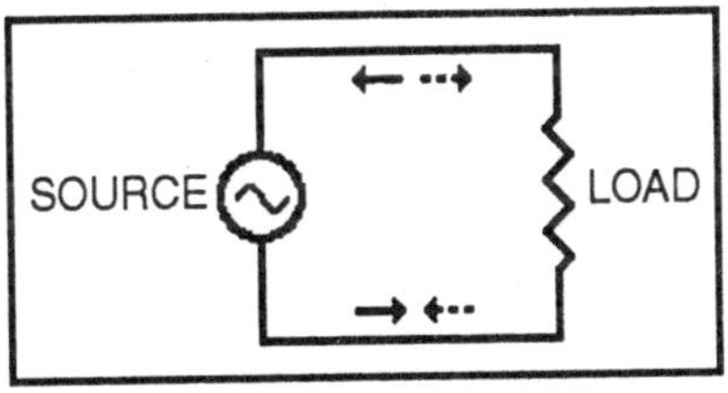

Now consider the same circuit with a single change, as shown in the second figure to the right. This time, the energy source is constantly changing. It begins by building up a voltage which is positive on top and negative on the bottom, and therefore pushes electrons through the circuit in the direction shown by the solid arrows.

However, then the source voltage starts to fall off, and eventually reverse polarity. Now current will still flow through the circuit, but this time in the direction shown by the dotted arrows.

This cycle repeats itself endlessly and as a result the current through the circuit reverses direction repeatedly. This is known as an alternating current.

This kind of reversal makes no difference to some kinds of loads. For example, the light bulbs in your home don't care which way current flows through them. When you close the circuit by turning on the light switch, the light turns on without regard for the direction of current flow.

Of course, there are some kinds of loads that require current to flow in only one direction. In such cases, we often need to convert alternating current such as the power provided at your wall socket to direct current for use by the load. There are several ways to accomplish this and we will explore some of them.

Properties of Alternating Current

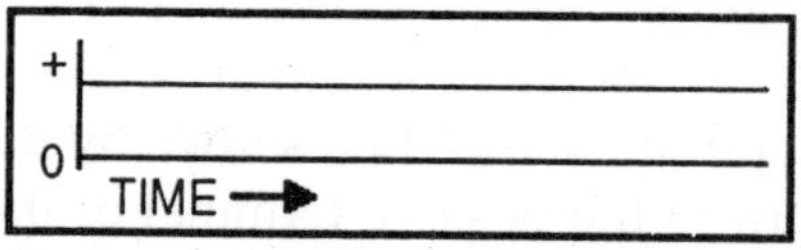

A DC power source, such as a battery, outputs a constant voltage over time, as depicted in the top figure to the right. Of course, once the chemicals in the battery have completed their reaction, the battery will be exhausted and cannot develop any output voltage. But until that happens, the output voltage will remain essentially constant. The same is true for any other source of DC electricity: the output voltage remains constant over time.

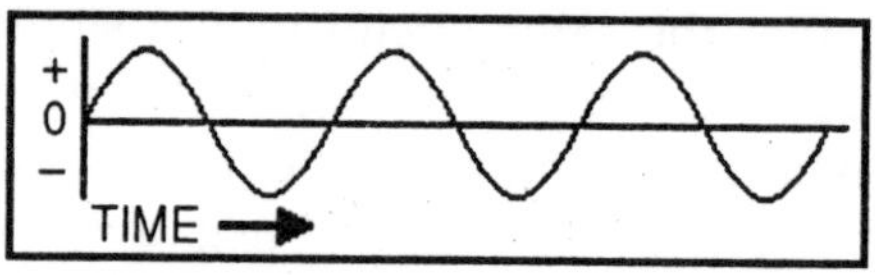

By contrast, an AC source of electrical power changes constantly in amplitude and regularly changes polarity, as shown in the second figure to the right. The changes are smooth and regular, endlessly repeating in a succession of identical cycles, and form a sine wave as depicted here.

Because the changes are so regular, alternating voltage and current have a number of properties associated with any such waveform.

These basic properties include the following list:

- *Frequency*: One of the most important properties of any regular waveform identifies the number of complete cycles it goes through in a fixed period of time. For standard measurements, the period of time is one second, so the frequency of the wave is commonly measured in cycles per second (cycles/sec) and, in normal usage, is expressed in units of Hertz (Hz). It is represented in mathematical equations by the letter'f. 'In North America (primarily the US and Canada), the AC power system operates at a frequency of 60 Hz. In Europe, including the UK, Ireland, and Scotland, the power system operates at a frequency of 50 Hz.
- *Period*: Sometimes we need to know the amount of time required to complete one cycle of the waveform, rather than the number of cycles per second of time. This is logically the reciprocal of frequency. Thus, period is the time duration of one cycle of the waveform, and is measured in seconds/cycle. AC power at 50 Hz will have a period of 1/50 = 0.02 seconds/cycle. A 60 Hz power system has a period of 1/60 = 0.016667 seconds/cycle. These are often expressed as 20 ms/cycle or 16.6667 ms/cycle, where 1 ms is 1 millisecond = 0.001 second (1/1000 of a second).

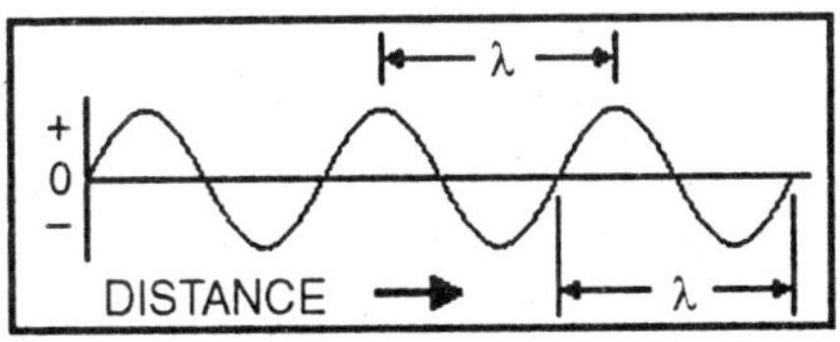

- *Wavelength*: Because an AC wave moves physically as well as changing in time, sometimes we need to know how far it moves in one cycle of the wave, rather than how long that cycle takes to complete. This of course depends on how fast the wave is moving as well. Electrical signals travel through their wires at nearly the speed of light, which is very nearly 3×10^8 metres/second and is represented mathematically by the letter 'c'. Since we already know the frequency of the wave in Hz or cycles/second, we can perform the division of c/f to obtain a result in units of metres/cycle, which is what we want. The Greek letter λ (lambda) is used to represent wavelength in mathematical expressions. Thus, $\lambda = c/f$. As shown in the figure to the right wavelength can be measured from any part of one cycle to the equivalent point in the next cycle. Wavelength is very similar to period as discussed above, except that wavelength is measured in distance per cycle where period is measured in time per cycle.

- *Amplitude*: Another thing we have to know is just how positive or negative the voltage is, with respect to some selected neutral reference. With DC, this is easy, the voltage is constant at some measurable value. But AC is constantly changing and yet it still powers a load. Mathematically, the amplitude of a sine wave is the value of that sine wave at its peak. This is the maximum value, positive or negative, that it can attain. However, when we speak of an AC power system, it is more useful to refer to the effective voltage or current. This is the rating that would cause the same amount of work to be done (the same effect) as the same value of DC voltage or current would cause. We won't cover the mathematical derivations here; for the present, we'll simply note that for a sine wave, the effective voltage of the AC power system is 0.707 times the peak voltage. Thus, when we say that the AC line voltage in the US is 120 volts, we are referring to the voltage amplitude, but we are describing the effective voltage, not the peak voltage of nearly 170 volts. The effective voltage is also known as the rms voltage.

When we deal with AC power, the most important of these properties are frequency and amplitude, since some types of electrically powered equipment must be designed to match the frequency and voltage of the power lines. Period is sometimes a consideration, as we'll discover when we explore electronic power supplies. Wave length is not generally important in this context but becomes much more important when we start dealing with signals at considerably higher frequencies.

Using Alternating Current

Since some kinds of loads require DC to power them and others can easily operate on either AC or DC, the question naturally arises,"Why not dispense entirely with AC and just use DC for everything?" This question is augmented by the fact that in some ways AC is harder to handle as well as to use. Nevertheless, there is a very practical reason, which overrides all other considerations for a widely distributed power grid. It all boils down to a question of cost.

DC does get used in some local commercial applications. An excellent example of this is the electric trolley car and trolley bus system used in San Francisco for public transportation. Trolley cars are electric train cars with power supplied by an overhead wire. Trolley busses are like any other bus, except they are electrically powered and get their power from two overhead wires. In both cases, they operate on 600 volts DC, and the overhead wires span the city.

The drawback is that most of the electrical devices on each car or bus, including all the light bulbs inside, are quite standard and require 110 to 120 volts. At the same time, however, if we were to reduce the system voltage, we would have to increase the amount of current drawn by each car or bus in order to provide the same amount of power to it. (Power is equal to the product of the applied voltage and the resulting current: $P = I \times E$) But those overhead wires are not perfect conductors; they exhibit some resistance. They will absorb some energy from the electrical current and dissipate it as waste heat in accordance with Ohm's Law

$$(E = I \times R)$$

With a small amount of algebra, we can note that the lost power can be expressed as:

$$Plost = I^2R$$

Now, if we reduce the voltage by a factor of 5 (to 120 volts DC), we must increase the current by a factor of 5 to maintain the same power to the trolley car or bus. But lost power is a function of the square of the current, so we will lose not five times as much power in the resistance of the wires but twenty-five times as much power. To offset and minimize that loss, we would have to use much larger wires, and pay a high price for all that extra copper. A cheaper solution is to mount a motor-generator set in each trolley car and bus, using a 600 volt dc motor and a lower-voltage generator to power all the equipment aboard that car.

The same reality of Ohm's Law and resistive losses holds true in the country-wide power distribution system. We need to keep the voltage used in homes to a reasonable and relatively safe value, but at the same time we need to minimize resistive losses in the transmission wires without bank rupting ourselves buying heavy-gauge copper wire.

At the same time, we can't use motor-generator pairs all across the country; they would need constant service and would break down far too often. We need a system that allows us to raise the voltage (and thus reduce the current) for long-distance transmission, and then reduce the voltage again (to a safe value) for distribution to individual homes and businesses. And we need to do this without requiring any moving parts to break down or need servicing.

The answer is to use an AC power system and transformers. (We'll learn far more about transformers in a later page; for now, a transformer is an electrical component that can convert incoming AC power at one voltage to outgoing power at a different voltage, higher or lower, with only very slight losses.) Thus, we can generate electricity at a reasonable voltage for practical AC generators (sometimes called alternators), then use transformers to step that voltage up to very high levels for long-distance transmission and then use additional transformers to step that high voltage back down for local distribution to individual homes.

In practice, this is done in stages. The really high-voltage transmission lines hanging from long glass insulators on the arms of tall steel towers carry electricity cross-country at several hundred thousand volts. This is stepped down to about 22,000 volts for distribution to multiple neighbourhoods – these are the wires you see at the top of the telephone poles in many areas. Additional transformers mounted on some of these telephone poles step this voltage down again for distribution to several homes each.

The design of the system minimizes the overall cost by balancing the cost of transformers against the cost of heavier-gauge copper wire, as well as the cost of maintaining the system and repairing damage. This is how the cost of electricity delivered to your home or business is kept to a minimum, while maintaining a very high level of service.

Examining DC Circuits

When we examined DC circuit theory and discussed DC power losses above, we used capital letters to represent all quantities. This is standard nomenclature; capital letters are used to represent fixed static values. Thus, we use a capital I to represent DC current and a capital V to represent DC voltage.

By the same token, we use the capital letters R, L and C to represent the values of circuit components containing resistance, inductance, and capacitance, respectively. These are fixed values that are set at the time the component is manufactured.

On the other hand, AC circuits use constantly-changing voltages and currents. In addition, some circuits contain both AC and DC components, which must often be considered separately. Therefore, we typically use lower-case letters to designate instantaneous values of voltage, current and power. Thus, if you see an equation written as:

$$P = I \times E$$

You know it refers to DC power, current and voltage. *On the other hand, an equation written as:*

$$p = i \times e$$

Refers to the instantaneous power, current and voltage of some AC signal at some specified instant in time.

There will also be cases where we must refer to an overall AC signal rather than one instantaneous value from it. In such cases, the use of upper and lower-case letters may be adjusted so we don't have too many of one or the other in a single discussion. We will try to avoid unnecessary confusion by including subscripts in equations or specifying in the text exactly what we are discussing.

SINUSOIDAL

Electric power is generated, distributed and used as sinusoidal voltages and currents in a great variety of commercial and domestic applications. Further more, in the industrial world a vast number of small-, medium- or large-size ac power generators and loads are interlinked. Therefore, the design, operation, maintenance, and management of such systems very much depend on a good understanding of ac circuit theory.

The analysis of ac circuits involves the study of the behaviour of the system under both normal and abnormal operating conditions. However, this book is not intended to include abnormal conditions. Instead, it focuses on the foremost fundamental issues and normal conditions and presents visual aids and interactive tools. Additionally, it is assumed that a steady-state sinusoidal condition is reached, which means that all transient effects in ac circuits have disappeared.

EDUCATIONAL OBJECTIVES

The chapter develops appropriate relationships and visual aids for describing ac systems using ac voltage, ac current, impedance, ac power, and phasors. After completing this chapter, students should be able to

- Plot and interpret the characteristics of ac voltage, current and power waveforms.
- Understand the definitions of peak to peak; peak and rms values; and phase/line voltage and current, phase angle, power factor, complex impedance, phasor and base (per-unit) values in ac circuits.

- State the equations for series, parallel, and combination ac circuits that contain impedances and ac supplies, understand the effects on current caused by changes in impedances, and state the meaning of the term equivalent impedance.
- Analyse the sinusoidal steady-state behaviour of single- and three-phase ac circuits using phasors, and study the effect of resistive, inductive and capacitive loads in single-phase ac circuits.

AC FUNDAMENTALS

ANALYSIS OF SERIES CIRCUITS

Ohm's Law teaches us that Voltage is equal to the product of resistance and current. This is fine if we have a circuit as simple as a single light run by a battery. When two circuits A and B are joined in series, the operation will result in a graph as shown in figure.

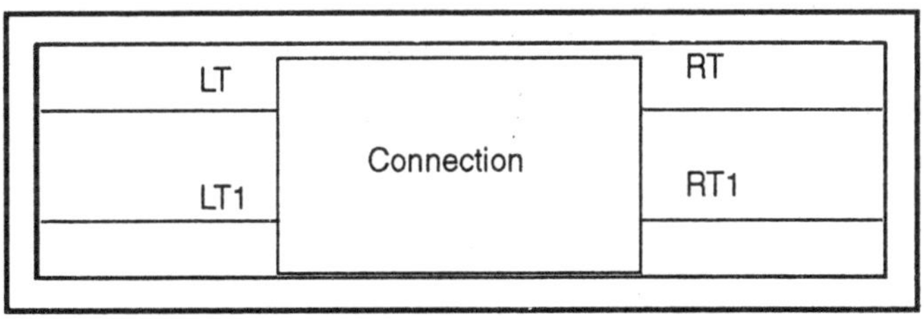

Fig. Joining Circuits in Series

But how often do we truly encounter a circuit which has only one resistance? It is actually quite seldom. Most circuits have many resistances in various combinations. So we must learn how to mathematically deal with all these resistances. Fortunately, combined resistances can only be configured in two ways:

Resistances in Series/Parallel

First, let us describe the difference between a series circuit and a parallel circuit. A series circuit is hooked together like a chain with each link connected to the link before it. If any given link in the chain is broken, the whole chain is broken and doesn't work.

If we think of our circuit as a water pipe, a resistor could be a valve in the pipe. A series circuit would have several valves in the same pipe. If we shut the water off at any one valve, the water flow through the entire pipe would be restricted. The flow or current is restricted by all the valves, but the valve with the most resistance is the one that decreases the flow the most.

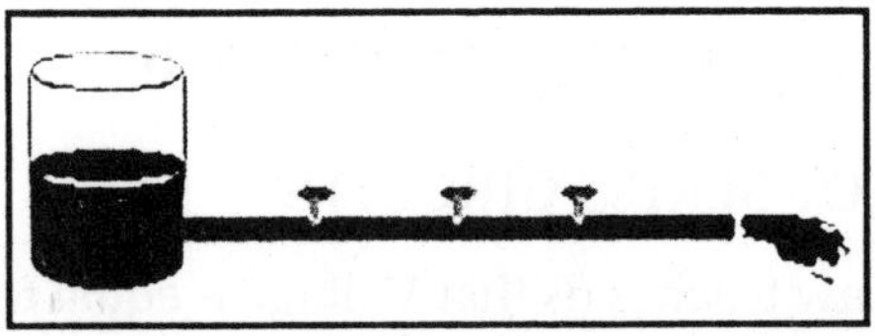

Now if we draw a schematic of resistors in series, it would look like the diagram at the left. This first diagram would be an exact representation of our water flow circuit. The tank provides a source of water, just as a battery supplies a source of electricity. The valves, which restrict the flow of water represent the resistors which restrict the flow of electric current.

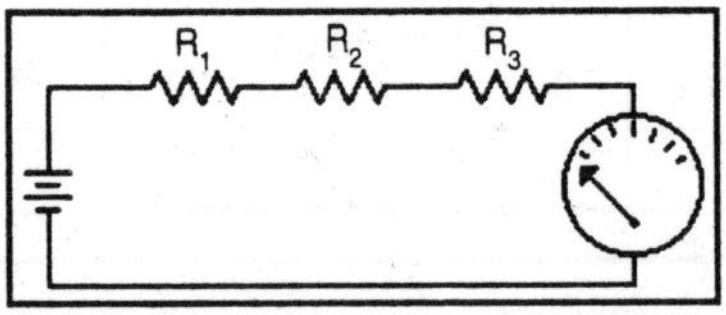

The diagram on the right shows the equivalent circuit, with only one resistor (Rt) taking the place of all 3 resistors in the left diagram. Recall that using an electrical equivalent circuit, that any circuit, no matter how complex, can be represented with a single voltage source and a single load. The question is, how do we determine the value of the total resistance when we have multiple resistors?

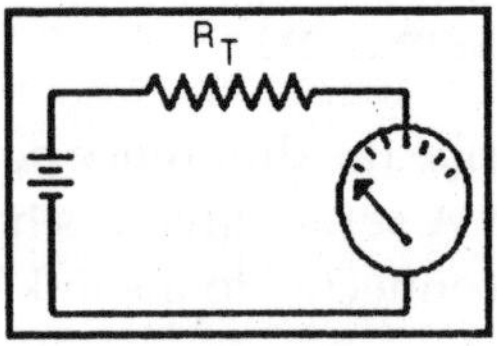

As said before... it's as easy as 2+2. If two or more resistors are connected end to end, as in a chain, we say that they are in series. To find out the total resistance of resistors in series, all we have to do is add up their individual values. It's that simple. If we have 3 resistors, (R1, R2, and R3) each with a value of 2W, the total resistance (RT) of the series circuit would be 2+2+2 or a total of 6W. Hence,

PARALLEL CIRCUITS

When two circuits A and B are joined parallel, the operation will result in a graph as shown in figure.

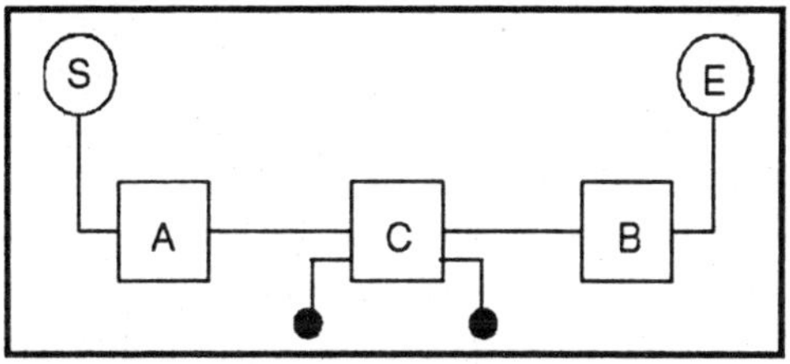

Fig. Joining Circuits in Parallel

In a parallel circuit the current divides and some flows through each component. The key to these circuits is that the voltage is the same in every element of the circuit and the total of the current through each of the components adds up to the total current from the battery. When you use a current metre it is always placed in series with the part of the circuit where you wish to measure the current.

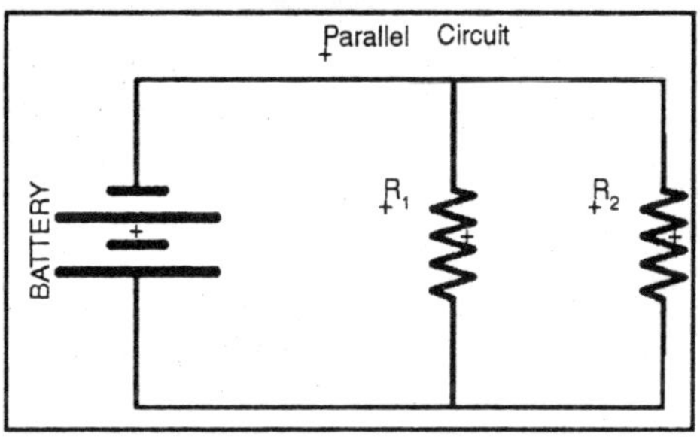

Circuit:

- R1 resistor 1 or any other 2 terminal component, capacitor, inductor, diode....

- R2 resistor 2 or any other 2 terminal component.....
- Battery a battery or other voltage source

Decscription

As we mentioned earlier, there were 2 types of circuits, series and parallel. We have already gone thru series. Now we are going to discuss the difference between a series and a parallel circuit.

In the series circuit, all the electricity followed the same path. In our waterflow representation, this meant that all the water flowed through 1 pipe. In parallel circuit, however, there are multiple paths that current can flow through.

Notice that in the picture to the left, that there are 3 different paths which the water can take. All 3 paths have the same incoming pressure, but the flow of some paths can be more restricted than in others.

Parallel circuits in electronics work on the same principle. While there may be multiple paths for the electricity to flow through, the electrical pressure (Voltage) remains the same through all paths.

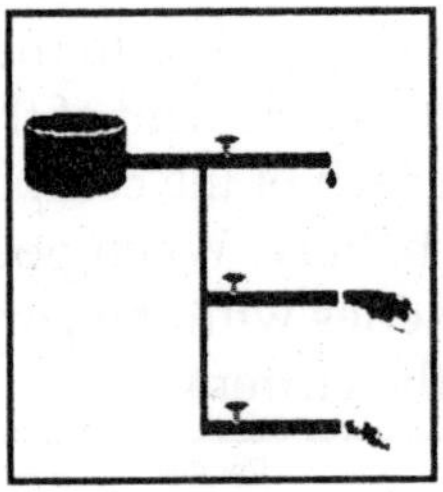

As you can see from the diagram, there are 4 metres placed in this circuit to measure the current. The first 3, (A_1, A_2 and A_3) measure only the current flowing through that individual leg of the circuit.

The 4th, at measures the Total current of the circuit. If you take the three individual currents and add them all together, they will equal the total current measured on the 4th metre. From this we can see that the current in a parallel circuit is additive.

$$I_1 + I_2 + I_X ... = I_{Total}$$

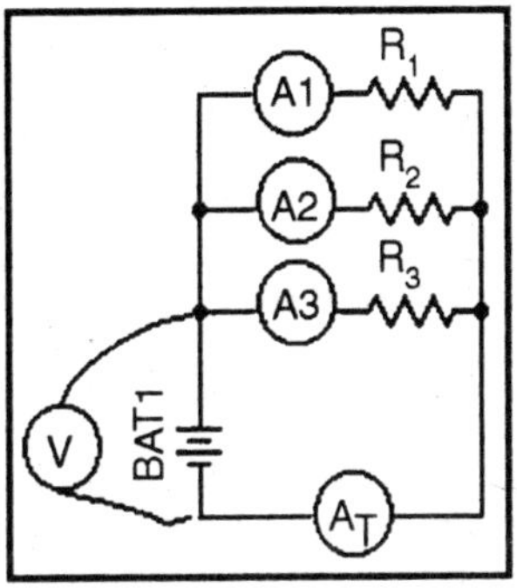

Resistance in a parallel circuit can be quite a bit trickier than in a series circuit. It is found by"Reciprocating the Sum of the Reciprocals".

Simple

Taking the reciprocal of a number means dividing"1" by that number. The reciprocal of 2 would be 1 divided by 2 or ½. Most modern calculators have a [1/X] button just for this purpose. So if you take the reciprocals of the values of all of the resistors, which would, of course, give you a bunch of fractions and add them all up, then reciprocate their sum, you would have the answer.

The formula would look something like this:

$$\frac{1}{\frac{1}{R_1} + \frac{1}{R_2} + \frac{1}{R_3} + \frac{1}{R_4} + \frac{1}{R_{X\ldots}} +}$$

Calculating Current, Voltage AND Power in Parallel Circuits

Let's begin with a very simple circuit: 2 resistors and 1 battery. You are given the following information about the circuit:

- R1 = 50 W
- R2 = 200 W
- A1 reads.2 Amps in current (I=.2)

Find:

- Total Voltage
- Total Circuit Resistance
- Total Current
- And Finally, the Current through A2

We know according to Ohm's law, that if we know the resistance and current, we can find the voltage.

- ER1= IR1 × R1.
- E =.2 × 50 = 10
- E = 10 Volts.

Now that we know that the voltage for the entire circuit is 10 volts, let's find the total Resistance.

First, we find the reciprocals of the individual resistances:

- R1= 50 ohms. 1/50 =.02
- R2= 200 ohms. 1/200 =.005

Now we add the two reciprocals together:

- .02 +.005 =.025

Finally we take the reciprocal of the sum:

- 1/.025 = 40 W

If the Total Voltage of the circuit is 10 Volts and the Total Resistance = 40 W then by using Ohms Law again we can find the total current.

- ITotal = ETotal/RTotal
- I = 10/40 = ¼ Ampere.

So far we know:

- R1 = 50 W
- R2 = 200 W
- A1 reads.2 Amps in current (I=.2)
- VTotal= 10
- RTotal= 40 and
- ITotal = ¼

Now we have at least 2 methods by which we can find the current through A2. We know that the Total current is the sum of all the individual leg currents, so if we subtract the current of A1 from the Total current we get this:

- ITotal - I1 = I2.25 -.2 = I2 =.05 Amperes.

The other method would be by using Ohms Law. We know the resistance of R2 = 200 W. We also know that the voltage across R2 = 10 Volts.

Hence:

- 10 Volts/200 W =.05 Amperes.
- Either way, our final result is A2 =.05 Amps

Chapter 4

Magnetic Circuits

MAGNETISM

Electromagnetism is produced when an electrical current flows through a simple conductor such as a piece of wire or cable. A small magnetic field is created around the conductor with the direction of this magnetic field with regards to its"North" and"South" poles being determined by the direction of the current flowing through the conductor.

Magnetism plays an important role in Electrical and Electronic Engineering because without it components such as relays, solenoids, inductors, chokes, coils, loudspeakers, motors, generators, transformers, and electricity metres etc, would not work if magnetism did not exist.

Then every coil of wire uses the effect of electromagnetism when an electrical current flows through it. But before we can look at Magnetism and especially Electromagnetism in more detail we need to remember back to our physics classes of how magnets and magnetism works.

DEFINITION

Magnets can be found in a natural state in the form of a magnetic ore, with the two main types being Magnetite also called"iron oxide", (FE_3O_4) and Lodestone, also called"leading stone". If these two natural magnets are suspended from a piece of string, they will take up a position inline with the earths magnetic field always pointing north. A good example of this effect is the needle of a compass. For most practical applications these natural occurring magnets can be disregarded as their magnetism is very low and because nowadays, manmade artificial magnets can be produced in many different shapes, sizes and magnetic strengths.

There are basically two forms of magnetism,"Permanent Magnets" and"Temporary Magnets", with the type being used dependant upon its application. There are many different types of materials available to make magnets such as iron, nickel, nickel alloys, chromium and cobalt and in their natural state some of these elements such as nickel and cobalt show very poor magnetic quantities on their own.

However, when mixed or"alloyed" together with other materials such as iron or aluminium peroxide they become very strong magnets producing unusual names such as"alcomax","hycomax","alni" and"alnico". Magnetic material in the non-magnetic state has its molecular structure in the form of loose magnetic chains or individual tiny magnets loosely arranged in a random pattern.

The overall effect of this type of arrangement results in zero or very weak magnetism as this haphazard arrangement of each molecular magnet tends to neutralise its neighbour. When the material is Magnetised this random arrangement of the molecules changes and the tiny molecular magnets become"lined-up" in such a way that they produce a series magnet arrangement..

MAGNETIC MOLECULE OF IRON AND MAGNET

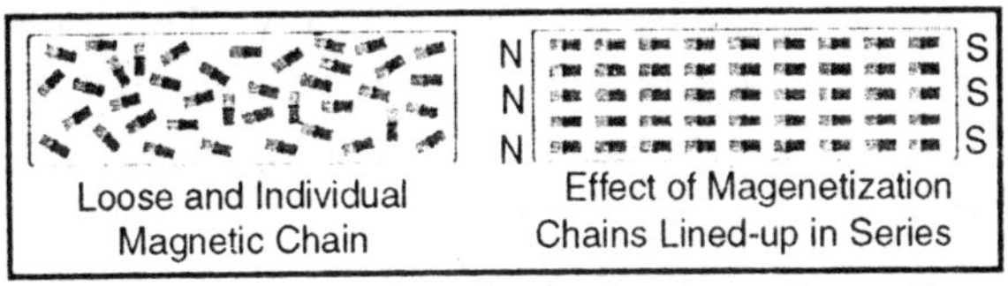

Weber's theory is based on the fact that all magnetic materials are composed of tiny magnets at a molecular level around the atoms, and a magnetised material will have most of its tiny magnets lined up in one direction only to produce a north pole in one direction and a south pole in the other direction.

Likewise, a material that has its tiny molecular magnets pointing in all directions will have its molecular magnets neutralised by its neighbouring magnet, thereby neutralising any magnetic effect. Once the magnetising force has been removed, the magnetism within the material will either remain or decay away quiet quickly depending on the magnetic material being used.

This ability of a material to retain its magnetism is called Retentivity and materials which are required to retain their magnetism will have a high retentivity and are used to make permanent magnets, while those materials required to loose their magnetism quickly such as soft iron cores for relays and solenoids will have a very low retentivity.

MAGNETIC FLUX

The magnetism in and around a magnetic circuit will have a definite chain producing an organised and balanced pattern of invisible lines around it and which are referred to as the"magnetic field" of the magnet. The shape of this magnetic field is more intense in some parts than others with these areas being called"poles" and each end of a magnet is a pole. They can be seen visually (called a vector field) by using iron fillings sprinkled onto a sheet of paper or by using a small compass to trace them out. Magnetic poles are always present in pairs, there is always a North-pole and there is always a South-pole.

Magnetic fields are always shown visually as lines of force that give a definite pole at each end of the material where the flux lines are more dense and concentrated. The lines which go to make up a magnetic field showing the direction and intensity are called Lines of Force or more commonly"Magnetic Flux" and are given the Greek symbol, Phi (F) as shown below.

LINES OF FORCE FROM A BAR MAGNETS MAGNETIC FIELD

As shown above, the magnetic field is strongest at the poles of the magnet where the lines of flux are more dense and the general direction for the magnetic flux flow is from the North (N) to the South (S) pole. Magnetic poles are always in pairs.

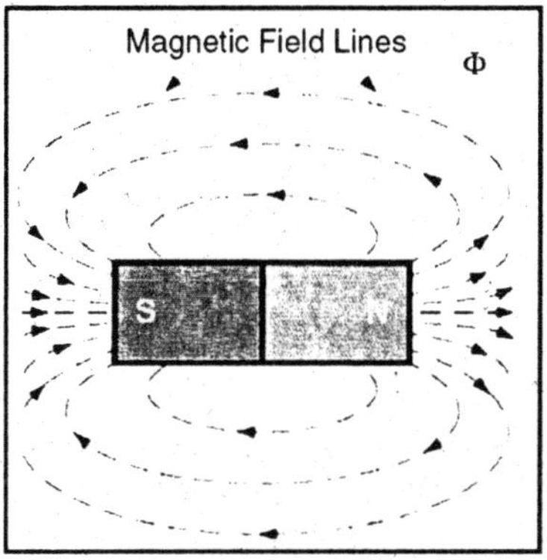

However, magnetic flux does not actually flow from the north to the south pole or flow anywhere for that matter as magnetic flux is a static region around a magnet in which the magnetic force exists. In other words magnetic flux does not flow or move it is just there!.

Some important facts emerge when plotting lines of force:

- Lines of force NEVER cross.
- Lines of force are CONTINUOUS.
- Lines of force always form individual CLOSED LOOPS around the magnet.
- Lines of force have a definite DIRECTION from North to South.
- Lines of force that are close together indicate a STRONG magnetic field.
- Lines of force that are farther apart indicate a WEAK magnetic field.

When two lines of force are brought close together the interaction between the two magnetic fields causes one of two things to occur:

- When adjacent poles are the same they REPEL each other.
- When adjacent poles are not the same they ATTRACT each other.

It can be remembered by the famous expression that"opposites attract" and this interaction of magnetic fields is easily demonstrated with iron fillings. The effect upon the magnetic fields of the various combinations of poles as like poles repel and unlike poles attract.

MAGNETIC FIELD OF LIKE AND UNLIKE POLES

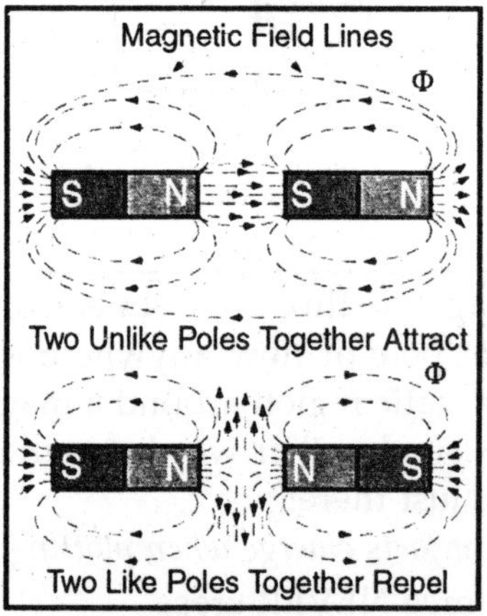

When plotting magnetic field lines with a compass it will be seen that the lines of force are produced in such a way as to give a definite pole at each end of the magnet where the lines of force leave the N-pole and re-enter at the S-pole. Magnetism can be destroyed by heating or hammering the magnetic material, but cannot be destroyed by simply breaking the magnet into two as each half will become a separate magnet with its own north and south poles. In order to make use of magnetism in electrical or electronic calculations, it is necessary to define the various aspects of magnetism.

MAGNITUDE OF MAGNETISM

We now know that the lines of force or more commonly the magnetic flux around a magnetic material is given the Greek symbol, Phi, (Φ) with the unit of flux being the Weber, (Wb) after Wilhelm Eduard Weber. But the number of lines of force within a given unit area is called the"Flux Density" and since flux (Φ) is measured in (Wb) and area (A) in metres squared, (m^2), flux density is therefore measured in Webers/ Metre2 or (Wb/m2) and is given the symbol B.

However, when referring to flux density in magnetism, flux density is given the unit of the Tesla after Nikola Tesla so therefore one Wb/m^2 is equal to one Tesla, $1Wb/m^2 = 1T$. Flux density is proportional to the lines of force and inversely proportional to area so we can define Flux Density as:

MAGNETIC FLUX DENSITY

$$\text{Magnetic Flux Density, (tesla)} = \frac{\text{Magnetic Flux, (weber)}}{\text{Area, (m2)}}$$

The symbol for magnetic flux density is B and the unit of magnetic flux density is the Tesla, T.

$$B = \frac{\Phi}{A} \text{ in Teslas}$$

It is important to remember that all calculations for flux density are done in the same units, e.g., flux in webers, area in m2 and flux density in teslas.

Example: The amount of flux present in a round magnetic bar was measured at 0.013 webers. If the material has a diametre of 12cm, calculate the flux density.

The cross sectional area of the magnetic material in m2 is given as:

$$\text{Diameter} = 12\text{cm}$$

$$\therefore \text{Area} = \pi r^2$$

$$A = 3.142 \times 0.06^2$$

$$A = 0.0113\text{m}^2$$

The magnetic flux is given as 0.013 webers, therefore the flux density can be calculated as:

$$B = \frac{\Phi}{A} = \frac{0.013}{0.0113} = 1.15T$$

So the flux density is calculated as 1.15 teslas.

When dealing with magnetism in electrical circuits it must be remembered that one tesla is the density of a magnetic field such that a conductor carrying 1 ampere at right angles to the magnetic field experiences a force of one newton-metre length on it and this will be demonstrated in the next tutorial about Electromagnetism.

MAGNETIC CIRCUIT

The term electrical circuit seems quite familiar and we often hear it in every day life. But do you know there is another type of circuit namely a magnetic circuit. If not, then here is an article which will explain some basic concepts of magnetic circuits.

THE MAGNETIC FIELD

Before going on to understand a magnetic circuit we must first understand what is meant by a magnetic field. Now you surely know that if you keep a magnet on the table and place a small iron object near it, it simply goes and sticks to the magnet.

The force with which the piece is pulled decreases as its distance from the magnet increases. This phenomenon can be explained if you consider that the space around the magnet is permeated with a type of force which influences the iron piece near it. This force field is known as the magnetic field which can be visualized to be in the form of magnetic lines of flux. These are just imagine lines and do not exist actually but can only be experienced.

Magnetic fields are created by electric currents in the space around where the currents flow. Currents which do not change with time (called direct currents or DC) make constant magnetic fields which we call DC fields. A gradual change in a direct current creates a corresponding gradual change in the DC field. By convention we refer to unchanging fields and fields which change in this slow non-periodic manner as DC fields.

Currents which change sign in a regular manner with time are called alternating currents or AC and give rise to corresponding AC magnetic fields. The most common AC fields are created by power lines and usually have fundamental frequencies of 50 or 60 Hz (referred to as"line" frequency) often with harmonics up to about 5 kHz. AC fields at other frequencies may be generated by rotating machines containing permanent magnets. Examples are magnetic stirrers and plasma etch machines which may make fields at about 0.3 Hz.

Table.The Units used to Measure Magnetic Fields are as Follows

SI unit of magnetic field strength	Amp/metre (A/m)
SI unit of magnetic flux density	Tesla (T)
CGS unit of magnetic flux density	Gauss (G)

The SI units are the modern units but the old CGS unit, the Gauss, is still so widely used that it has been kept in the Spicer Consulting systems. The old CGS unit of magnetic field strength, the Oersted, is now rarely used.

The Amp/metre unit is commonly used by the electricity supply industry as it relates directly to the currents which are generating the magnetic field. The Tesla and Gauss are units of the flux density created by the magnetic field and are the most common units used in measurement of fields. The relationship between the units (in air or space) is as follows...1 Amp/metre = 1.257 microTesla = 12.57 milliGauss

Electron Microscopes

Electron microscopes control the motion of their electrons using electric and magnetic fields inside the part of the instrument called the"column". Focussed ion beam systems are similar. The column usually comprises a stack of electron optical components,"lenses" to focus the electron beam and"deflectors" to position the beam. The column is mostly made of ferromagnetic iron which serves as the magnetic circuit for the lenses and the mechanical structure of the column. The magnetic fields inside some of the lenses can be as large as 10,000 Gauss (1.0 Tesla) i.e. 20,000 times the earth's field. The iron structure of the column serves as a partial barrier to magnetic fields entering the column from outside and the large internal lens fields leaking out.

Ambient magnetic fields penetrate the electron beam column from outside (to some extent) and add to the magnetic fields in the lenses and deflectors. AC fields, and DC fields which change during operation of the microscope, may degrade the performance of the microscope. Usually, DC fields penetrate the column more than AC fields. This is because AC fields are also attenuated by the mechanism called"eddy current shielding".

Changing DC fields, which cause a problem for electron microscopes, can be created by steel objects moving and locally distorting the earth's field. Examples include cupboard doors, chairs, motor vehicles and elevators (lifts). Large changing DC fields can be created by electric trams and commuter trains which draw direct current from an overhead wire, returning it through the rails. Superconducting magnets in research facilities are also not good news for electron microscopes because they can radiate large DC fields for tens of metres.

The earth's magnetic field also penetrates the electron beam column but, for most electron microscopes, it can be considered constant since it changes slowly compared with the time taken for a micrograph. Its effect is adjusted out with the instrument controls during routine operation of the microscope and the operator is generally unaware of its presence. For electron beam lithography machines used to write reticles for IC chips (or to write directly on a silicon wafer) the changes in the earth's magnetic field during the write time (several hours) may significantly degrade the accuracy of the written pattern. Some EBeam lithography machine manufacturers are now specifying maximum field changes of 0.1 mG. Since the earth's field changes much more than this on a typical day, magnetic field cancelling would be necessary for these machines to meet specification.

Electron microscopes form their image either by projection (TEMs) or by scanning to form a TV like image (SEMs, STEMs). The effect of DC or AC fields on projected images is to move the whole image and the result is loss of resolution on the image screen. The effect of DC or AC fields on scanned images is to move parts of the image. AC line fields typically produce wiggly edges on the image often called"edge tearing" or"ragging".

DC fields may cause discontinuities in the image and cause straight edges to look like mountain ranges. Some microscope manufacturers design their scanning sytems to operate synchronously with the 50 or 60 Hz line. This removes the edge tearing caused by the AC line fields but introduces distortion into the image (e.g. the image of a square grid is bent). However, distortion is generally less visible than edge tearing and may be acceptable for some applications.

Earth's Magnetic Field

The magnetic field of the Earth (geomagnetic field) resembles the field produced by a simple bar magnet. Such a field is called a dipole field because it has two poles, located at either end of the magnet, where the strength of the field is maximum. At the midpoint between the poles the strength is half of its value at the poles. The magnetic field is often visualized in terms of magnetic field lines, or lines of force, that leave one end of the magnet, called the north pole, arc through space, and re-enter the magnet at the other end, the south pole.

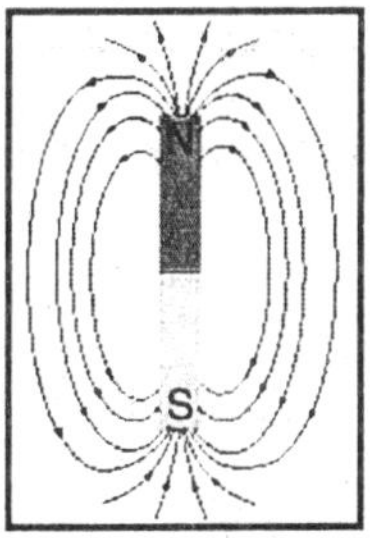

Fig. Magnetic Field Lines

If we could place a bar magnet inside the Earth, inclined approximately 11° to the rotational axis and offset about 550 km from the Earth's centre we could account for 90% of the observed magnetic field. We could account for the remaining 10% by placing smaller bar magnets at strategic locations around the primary magnet.

Although this simple description provides a way of conceptualizing the magnetic field it does not represent what really goes inside the Earth. Nor can it be used to calculate the strength and direction of the magnetic field. Mathematically, the Earth's magnetic field is normally described by a spherical harmonic expansion, a series of special spherical functions of latitude/longitude and their associated coefficients. Each group of functions describes a particular field pattern: the first three terms describe the field of a dipole; the next 5 terms describe a quadrapole field; the following 7 terms describe an octopole field. By adjusting the relative size of each of these component parts the shape of the actual field can be closely approximated.

Such mathematical descriptions of the magnetic field are called magnetic reference field models. The most widely used reference field model is the International Geomagnetic Reference Field (IGRF). The IGRF is produced every five years under the auspices of the International Association of Geomagnetism and Aeronomy (IAGA). Charts of the magnetic declination and total intensity, the two most widely used magnetic field elements, are shown in the accompanying figures.

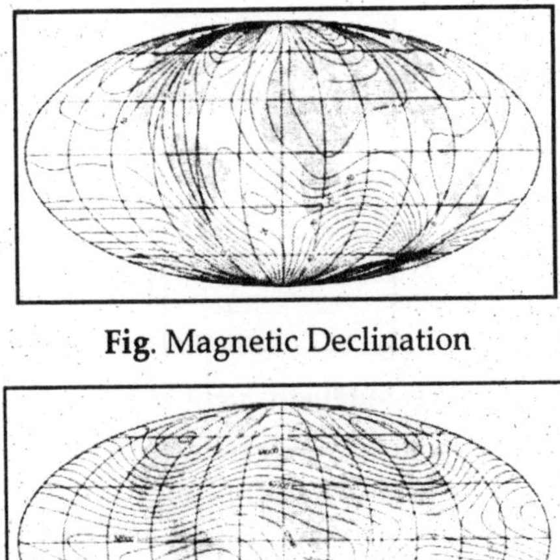

Fig. Magnetic Declination

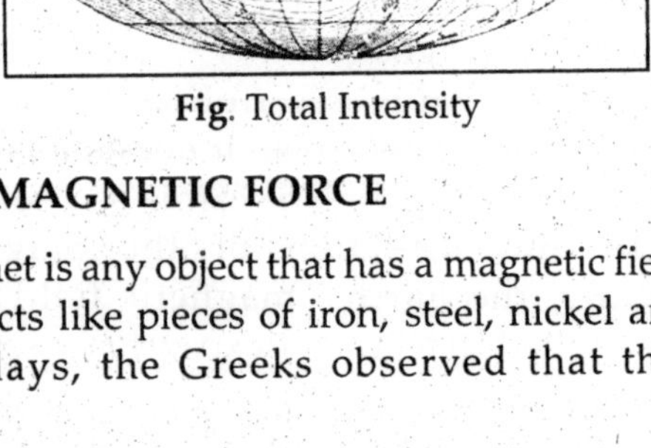

Fig. Total Intensity

LINES OF MAGNETIC FORCE

A magnet is any object that has a magnetic field. It attracts ferrous objects like pieces of iron, steel, nickel and cobalt. In the early days, the Greeks observed that the naturally occurring 'lodestone' attracted iron pieces. From that day onwards began the journey into the discovery of magnets.

These days magnets are made artificially in various shapes and sizes depending on their use. One of the most common magnets - the bar magnet - is a long, rectangular bar of uniform cross-section that attracts pieces of ferrous objects. The magnetic compass needle is also commonly used. The compass needle is a tiny magnet which is free to move horizontally on a pivot. One end of the compass needle points in the North direction and the other end points in the South direction.

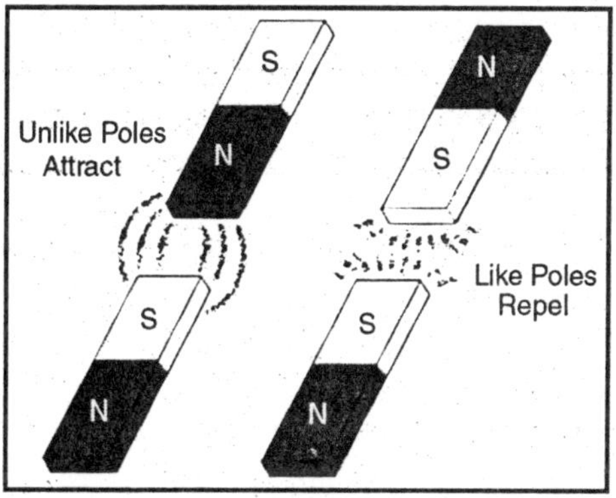

Fig. The end of a Freely Pivoted Magnet will always Point in the North-South Direction

The end that points in the North is called the North Pole of the magnet and the end that points South is called the South Pole of the magnet. It has been proven by experiments that like magnetic poles repel each other whereas unlike poles attract each other.

Magnetic Fields

What is a magnetic field? The space surrounding a magnet, in which magnetic force is exerted, is called a magnetic field. If a bar magnet is placed in such a field, it will experience magnetic forces. However, the field will continue to exist even if the magnet is removed. The direction of magnetic field at a point is the direction of the resultant force acting on a hypothetical North Pole placed at that point.

When current flows in a wire, a magnetic field is created around the wire. From this it has been inferred that magnetic fields are produced by the motion of electrical charges. A magnetic field of a bar magnet thus results from the motion of negatively charged electrons in the magnet.

Magnetic Lines Of Force

Just as an electric field is described by drawing the electric lines of force, in the same way, a magnetic field is described by drawing the magnetic lines of force. When a small north magnetic pole is placed in the magnetic field created by a magnet, it will experience a force.

And if the North Pole is free, it will move under the influence of magnetic field. The path traced by a North magnetic pole free to move under the influence of a magnetic field is called a magnetic line of force. In other words, the magnetic lines of force are the lines drawn in a magnetic field along which a north magnetic pole would move.

The direction of a magnetic line of force at any point gives the direction of the magnetic force on a north pole placed at that point.

Since the direction of magnetic line of force is the direction of force on a North Pole, so the magnetic lines of force always begin on the N-pole of a magnet and end on the S-pole of the magnet. A small magnetic compass when moved along a line of force always sets itself along the line tangential to it. So, a line drawn from the South Pole of the compass to its North Pole indicates the direction of the magnetic field.

Properties of the Magnetic Lines of Force

- The magnetic lines of force originate from the North Pole of a magnet and end at its South Pole.
- The magnetic lines of force come closer to one another near the poles of a magnet but they are widely separated at other places.
- The magnetic lines of force do not intersect (or cross) one another.
- When a magnetic compass is placed at different

points on a magnetic line of force, it aligns itself along the tangent to the line of force at that point.

These are just some of the basic concepts of magnetism. One cannot possibly grasp the depth and appreciate the versatility of magnets without reading more about the uses of magnets, the Earth as a huge magnet and electromagnetism among other things.

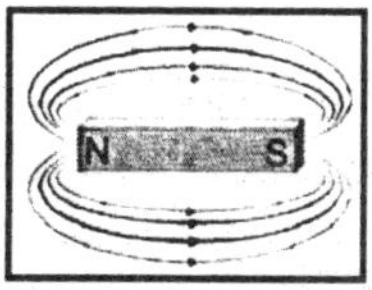

Description

To further describe and work with magnet phenomena, lines are used to represent the force existing in the area surrounding a magnet (refer to fig.). These lines, called MAGNETIC LINES OF FORCE, do not actually exist but are imaginary lines used to illustrate and describe the pattern of the magnetic field. The magnetic lines of force are assumed to emanate from the north pole of a magnet, pass through surrounding space, and enter the south pole. The lines of force then travel inside the magnet from the south pole to the north pole, thus completing a closed loop.

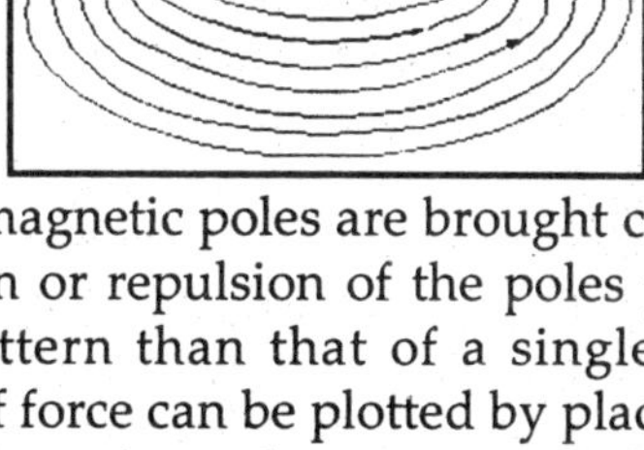

When two magnetic poles are brought close together, the mutual attraction or repulsion of the poles produces a more complicated pattern than that of a single magnet. These magnetic lines of force can be plotted by placing a compass at various points throughout the magnetic field, or they can be roughly illustrated by the use of iron filings as before. A diagram of magnetic poles placed close together is shown in figure below.

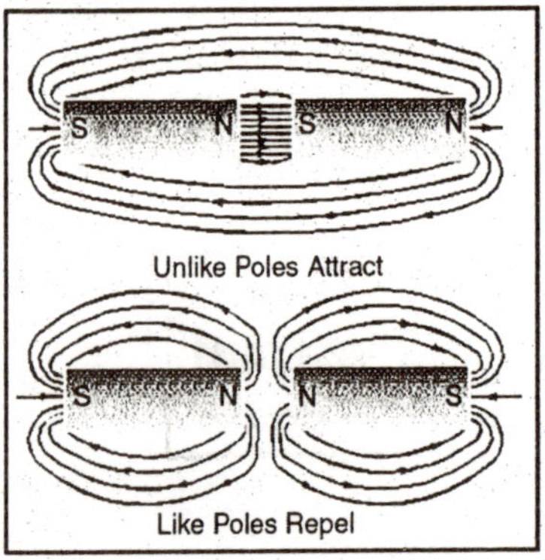

Although magnetic lines of force are imaginary, a simplified version of many magnetic phenomena can be explained by assuming the magnetic lines to have certain real properties. The lines of force can be compared to rubber bands which stretch outward when a force is exerted upon them and contract when the force is removed.

The characteristics of magnetic lines of force can be described as follows:

- Magnetic lines of force are continuous and will always form closed loops.
- Magnetic lines of force will never cross one another.
- Parallel magnetic lines of force traveling in the same direction repel one another. Parallel magnetic lines of force traveling in opposite directions tend to unite with each other and form into single lines traveling in a direction determined by the magnetic poles creating the lines of force.
- Magnetic lines of force tend to shorten themselves. Therefore, the magnetic lines of force existing between two unlike poles cause the poles to be pulled together.
- Magnetic lines of force pass through all materials, both magnetic and nonmagnetic.
- Magnetic lines of force always enter or leave a magnetic material at right angles to the surface.

Magnetic Shielding

There is no known INSULATOR for magnetic flux. If a nonmagnetic material is placed in a magnetic field, there is no appreciable change in flux - that is, the flux penetrates the nonmagnetic material. For example, a glass plate placed between the poles of a horseshoe magnet will have no appreciable effect on the field although glass itself is a good insulator in an electric circuit. If a magnetic material (for example, soft iron) is placed in a magnetic field, the flux may be redirected to take advantage of the greater permeability of the magnetic material, as shown in figure below. Permeability, as discussed earlier, is the quality of a substance which determines the ease with which it can be magnetized.

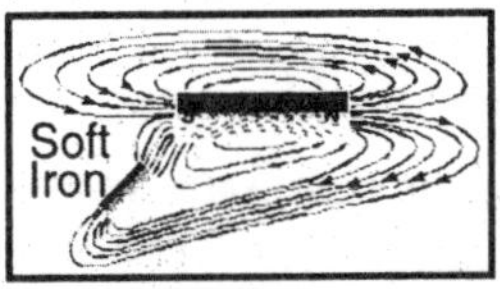

The sensitive mechanisms of electric instruments and metres can be influenced by stray magnetic fields which will cause errors in their readings. Because instrument mechanisms cannot be insulated against magnetic flux, it is necessary to employ some means of directing the flux around the instrument. This is accomplished by placing a soft-iron case, called a MAGNETIC SCREEN or SHIELD, about the instrument. Because the flux is established more readily through the iron (even though the path is longer) than through the air inside the case, the instrument is effectively shielded, as shown by the watch and soft-iron shield in figure below.

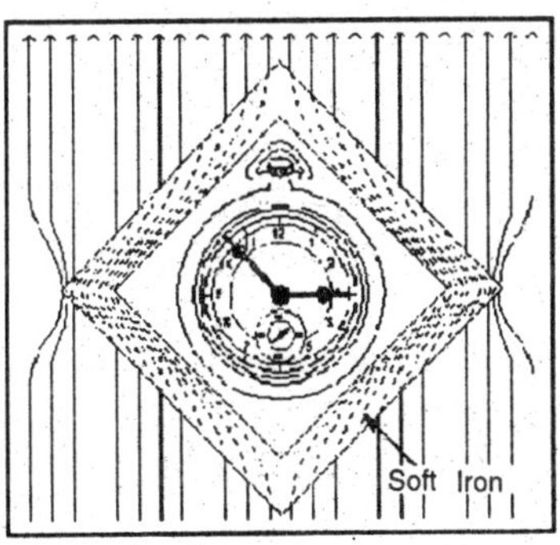

FEATURES OF LINES OF FLUX

There are several characteristics of lines of force which are as follows:

- *Direction*: Any line of flux always points towards the north seeking pole of a magnet provided it is not under the influence of another magnetic medium except the magnetic field of the earth.
- *Closed Line*: All lines of flux form a closed loop which emerge from the North Pole into the South Pole of the magnet Independent
- *Region*: Lines of flux never intersect each other. Repulsion: lines of flux which are in the same direction always repel each other.

THE MAGNETIC CIRCUIT

A magnetic circuit is made up of one or more closed loop paths containing a magnetic flux. The flux is usually generated by permanent magnets or electromagnets and confined to the path by magnetic cores consisting of ferromagnetic materials like iron, although there may be air gaps or other materials in the path.

Magnetic circuits are employed to efficiently channel magnetic fields in many devices such as electric motors, generators, transformers, relays, lifting electromagnets, SQUIDs, galvanometres, and magnetic recording heads.

The concept of a"magnetic circuit" exploits a one-to-one correspondence between the equations of the magnetic field in a non-hysteretic material to that of an electrical circuit. Using this concept the magnetic fields of complex devices such as transformers can be quickly solved using the methods and techniques developed for electrical circuits.

Some examples of magnetic circuits are:

- Horseshoe magnet with iron keeper (low-reluctance circuit)
- Horseshoe magnet with no keeper (high-reluctance circuit)
- Electric motor (variable-reluctance circuit)

The above discussion about magnetic field and lines of flux now leads us to the definition of a magnetic circuit. A magnetic circuit can be described as a complete closed path followed by any group of lines of magnetic flux.

A magnetic circuit is quite similar to an electrical circuit in terms of various parametres which are associated with the magnetic circuit, some of which are described as follows.

Magneto-motive Force: Just as electromotive force or emf makes electric current flow, similarly magneto-motive force or mmf makes the magnetic flux possible in a magnetic circuit.

The units of mmf are ampere-turns, where turns signify the number of electric coil turns which is generating that mmf. Since number of terms is a dimensionless quantity, effectively the units of mmf are same as that of current, namely amperes.

Magnetic Field Strength: it refers to the mmf that exists per unit length of a magnetic circuit provided the circuit is homogeneous and of a uniform cross sectional area. In the earlier days it was also known as magnetizing force but this term is obsolete now. The units of magnetic field strength are amperes/metre.

Magnetic Circuit Reluctance

The magnetic system's core can be represented as a magnetic circuit carrying the magnetic flux. The magnetic circuit reluctance can be estimated using the simulation software.

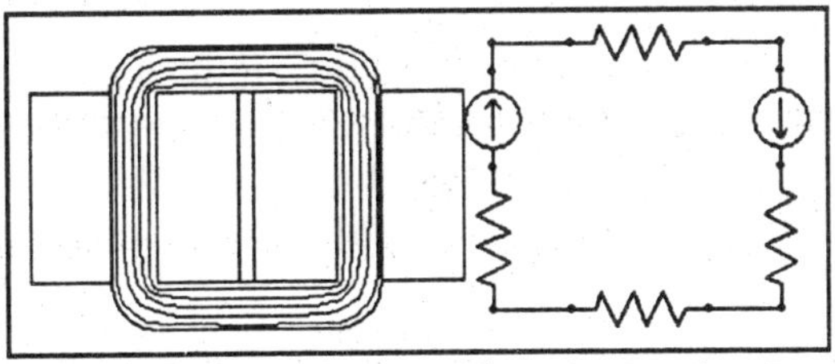

Magnetic circuit reluctance may be calculated as $R_m = U_m$/Flux, where R_m - magnetic circuit resistance

U_m = Integral(HdL) - magnetic voltage drop,

Flux - magnetic flux across the cross-section.

In case of uniform magnetic flux distribution the magnetic circuit reluctance can be calculated as,

$$R_m = (1/\mu\mu_0) \cdot L/S$$

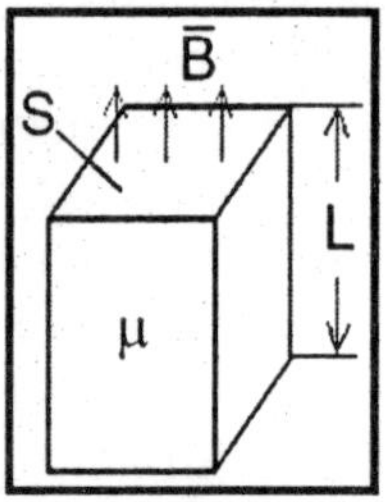

The magnetic circuit analysis can be performed by the same methods as the electric circuit analysis. This approach is suitable for simple geometry and uniform flux distribution.

For complex core construction and non-linear materials the FEA approach is more appropriate. Depending on the problem formulation, you can use DC, AC or Transient Magnetics module of QuickField to calculate the magnetic circuit reluctance of the core.

ELECTRIC AND MAGNETIC CIRCUITS

Analogies between magnetic circuits and electrical circuits are easy to be made:

- The magnetic flux ψ which flows through a magnetic circuit, corresponds to the electrical current/which flows through an electrical circuit;
- The magneto motive force F, corresponds to the electromotive force U;
- Reluctance R of a magnetic conductor by ℓlenght, S section and μpermeability, corresponds to resistance R of an electrical conductor by ℓlenght, S section and σconductivity; we can write,

$$R = \frac{\ell}{\mu.S} \text{ si } R = \frac{\ell}{\sigma.S};$$

- Finally Hopkinson's Law $F = R.\psi$ Corresponds to Ohm's Law $U = R.I$.

We can also define the permeance of a magnetic circuit $P = 1/R$, which corresponds to $G = 1/R$ conductance of an electrical circuit.

Table. Analogy Magnetic Circuits/Electrical Circuits

Magnetic circuit	Electrical circuit
Flux (ψ)	Current (I)
Magneto motive force(F)	Electromotive force (U)
Reluctance (R)	Resistance (R)
Permeance (P)	Conductance (G)
Hopkison's Law (F = R.)	Ohm's Law (U = R.I)

MAGNETIC LEAKAGE

DEFINITION

Magnetic flux leakage (MFL) is a magnetic method of nondestructive testing that is used to detect corrosion and pitting in steel structures, most commonly pipelines and storage tanks. The basic principle is that a powerful magnet is used to magnetize the steel.

At areas where there is corrosion or missing metal, the magnetic field"leaks" from the steel. In an MFL tool, a magnetic detector is placed between the poles of the magnet to detect the leakage field. Analysts interpret the chart recording of the leakage field to identify damaged areas and hopefully to estimate the depth of metal loss. This article currently focuses mainly on the pipeline application of MFL, but links to tank floor examination are provided at the end.

There are many methods of assessing the integrity of a pipeline. In-line-Inspection (ILI) tools are built to travel inside a pipeline and collect data as they go. The type of ILI we are interested in here, and the one that has been in use the longest for pipeline inspection, is the magnetic flux leakage inline inspection tool (MFL-ILI). MFL-ILIs detect and assess areas where the pipe wall may be damaged by corrosion.

The more advanced versions are referred to as"high-resolution" because they have a large number of sensors. The high-resolution MFL-ILIs allow more reliable and accurate identification of anomalies in a pipeline, thus, minimizing the need for expensive verification excavations (i.e. digging up the pipe to verify what the problem is). Accurate assessment of pipeline anomalies can improve the decision making process within an Integrity Management Programme and excavation programme can then focus on required repairs instead of calibration or exploratory digs. Utilizing the information from an MFL ILI inspection is not only cost effective but, as well, can also prove to be an extremely valuable building block of a Pipeline Integrity Management Programme.

The reliable supply and transportation of product in a safe and cost-effective manner is a primary goal of most pipeline operating companies and managing the integrity of the pipeline is paramount in maintaining this objective. In-line-inspection programmes are one of the most effective means of obtaining data that can be used as a fundamental base for an Integrity Management Programme.

There are many types of ILI tools that detect various pipeline defects, but high-resolution MFL tools are becoming more prevalent as its applications are surpassing those to which it was originally designed. Originally designed for detecting areas of metal loss, the modern High Resolution MFL tool is proving to be able to accurately assess the severity of corrosion features, define dents, wrinkles, buckles, and, in some cases, cracks. Having a device that can perform simultaneous tasks reliably is more efficient and ultimately provides cost saving benefits.

TOOL

In the field, a device that travels inside a pipeline to clean or inspect it is typically known as a"PIG". There is no history to back up this term as an acronym as it was first derived from the squealing pig-like sound made as the tool passed by. In some countries a pig is known as a"Diablo," literally translated to mean"the Devil" relating to the shuddering sound the tool would make as it passed beneath people's feet.

The pigs are built to match the diametre of a pipeline and use the very product being carried to end users to transport them. Pigs have been used in pipelines for many years and have many uses. Some separate one product from another, some clean and some inspect. An MFL tool is known as an"intelligent" or"smart" inspection pig because it contains electronics and collects data real-time while travelling through the pipeline. Sophisticated electronics on board allow this tool to accurately detect features as small as 1 cm by 1 cm.

Typically, an MFL tool consists of two or more bodies. One body is the magnetizer with the magnets and sensors and the other bodies contain the electronics and batteries. The magnetizer body houses the sensors that are located between powerful"rare-earth" magnets. The magnets are mounted between the brushes and tool body to create a magnetic circuit along with the pipe wall. As the tool travels along the pipe, the sensors detect interruptions in the magnetic circuit.

Interruptions are typically caused by metal loss and which in most cases is corrosion. Mechanical damage such as shovel gouges can also be detected. The metal loss in a magnetic circuit is analogous to a rock in a stream. Magnetism needs metal to flow and in the absence of it, the flow of magnetism will go around, over or under to maintain its relative path from one magnet to another, similar to the flow of water around a rock in a stream. The sensors detect the changes in the magnetic field in the three directions (axial,radial, or circumferential) to characterize the anomaly. An MFL tool can take sensor readings based on either the distance the tool travels or on increments of time. The choice depends on many factors such as the length of the run, the speed that the tool intends to travel, and the number of stops or outages that the tool may experience.

The second body is called an Electronics Can. This section can be split into a number of bodies depending on the size of the tool. This can, as the name suggests, contains the electronics or"brains" of the smart pig. The Electronics Can also contains the batteries and is some cases an IMU (Inertial Measurement Unit) to tie location information to GPS coordinates. On the very rear of the tool are odometre wheels that travel along the inside of the pipeline to measure the distance and speed of the tool.

PRINCIPLE

As an MFL tool navigates the pipeline a magnetic circuit

is created between the pipewall and the tool. Brushes typically act as a transmitter of magnetic flux from the tool into the pipewall, and as the magnets are oriented in opposing directions, a flow of flux is created in an elliptical pattern. High Field MFL tools saturate the pipewall with magnetic flux until the pipewall can no longer hold any more flux. The remaining flux leaks out of the pipewall and strategically placed tri-axial Hall effect sensor heads can accurately measure the three dimensional vector of the leakage field.

Given the fact that magnetic flux leakage is a vector quantity and that a hall sensor can only measure in one direction, three sensors must be oriented within a sensor head to accurately measure the axial, radial and circumferential components of an MFL signal. The axial component of the vector signal is measured by a sensor mounted orthogonal to the axis of the pipe, and the radial sensor is mounted to measure the strength of the flux that leaks out of the pipe.

The circumferential component of the vector signal can be measured by mounting a sensor perpendicular to this field. Earlier MFL tools recorded only the axial component but high-resolution tools typically measure all three components. To determine if metal loss is occurring on the internal or external surface of a pipe, a separate eddy current sensor is utilized to indicate wall surface location of the anomaly. The unit of measure when sensing an MFL signal is the gauss or the tesla and generally speaking, the larger the change in the detected magnetic field, the larger the anomaly.

ANALYSIS

The primary purpose of an MFL tool is to detect corrosion in a pipeline. To more accurately predict the dimensions (length, width and depth) of a corrosion feature, extensive testing is performed before the tool enters an operational pipeline. Using a known collection of measured defects, tools can be trained and tested to accurately interpret MFL signals. Defects can be simulated using a variety of methods.

Creating and therefore knowing the actual dimensions of a feature makes it relatively easy to make simple correlations of signals to actual anomalies found in a pipeline. When signals in an actual pipeline inspection have similar characteristics to the signals found during testing it is logical to assume that the features would be similar. The algorithms and neural nets designed for calculating the dimensions of a corrosion feature are complicated and often they are closely guarded trade secrets. An anomaly is often reported in a simplified fashion as a cubic feature with an estimated length, width and depth. In this way, the effective area of metal loss can be calculated and used in acknowledged formulas to predict the estimated burst pressure of the pipe due to the detected anomaly.

Another important factor in the ongoing improvement of sizing algorithms is customer feedback to the ILI vendors. Every anomaly in a pipeline is unique and it is impossible to replicate in the shop what exists in all cases in the field. Open lines of communication usually exist between the inspection companies and the pipeline operators as to what was reported and what was actually visually observed in an excavation.

After an inspection, the collected data is downloaded and compiled so that an analyst is able to accurately interpret the collected signals. Most pipeline inspection companies have proprietary software designed to view their own tool's collected data. The three components of the MFL vector field are viewed independently and collectively to identify and classify corrosion features. Metal loss features have unique signals that analysts are trained to identify.

B-H CURVE

Magnetic Hysteresis relates to the magnetization properties of a material in which it becomes magnetized and then de-magnetized. We know that the magnetic flux generated by an electromagnetic coil is the amount of magnetic field or lines of force produced within a given area and that it is more commonly called"Flux Density".

Given the symbol B with the unit of flux density being the Tesla, T. We also know from the previous tutorials that the magnetic strength of an electromagnet depends upon the number of turns of the coil, the current flowing through the coil or the type of core material being used, and if we increase either the current or the number of turns we can increase the magnetic field strength, symbol H.

In the previous tutorial about Electromagnets the relative permeability, symbol μr was defined as the product of μ (absolute permeability) and μ_0 the permeability of free space and was given as a constant. However, the relationship between B and H can be defined by the fact that the relative permeability, μr is not a constant but a function of the magnetic field intensity thereby giving magnetic flux density as: B = μH.

So for ferromagnetic materials the ratio of flux density to field strength (B/H) is not constant but varies with flux density. However, for air cored coils or any non-magnetic medium core such as woods or plastics, this ratio can be considered as a constant and this constant is known as μ_0, the permeability of free space. By plotting values of flux density, (B) against field strength, (H) we can produce a set of curves called Magnetization Curves, Magnetic Hysteresis Curves or more commonly B-H Curves for individual types of core materials as shown below.

MAGNETIZATION OR B-H CURVE

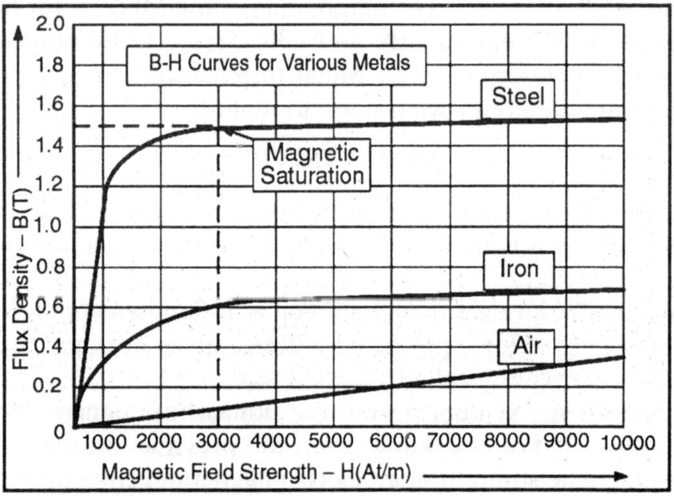

The set of curves above represents an example of the

relationship between B and H for soft-iron and steel cores but every type of core material will have its own set of curves. You may notice that the flux density increases in proportion to the field strength until it reaches a point were it can not increase any more becoming almost level and constant as the field strength continues to increase. This is because there is a limit to the amount of flux density that can be generated by the core. The condition were the flux density reaches its limit is called Magnetic Saturation also known as Saturation of the Core and in our simple example above the saturation point of the steel curve begins at about 3000 ampere-turns per metre.

Saturation occurs because as we remember from the previous Magnetism tutorial which included Weber's theory, the random haphazard arrangement of the molecule structure within the core material changes as the tiny molecular magnets within the material become"lined-up". As the magnetic field strength, (H) increases these molecular magnets become more and more aligned until they reach perfect alignment producing maximum flux density and any increase in the magnetic field strength due to an increase in the electrical current flowing through the coil will have little or no effect.

RETENTIVITY

Lets assume that we have an electromagnetic coil with a high field strength due to the current flowing through it, and that the ferromagnetic core material has reached its saturation point, maximum flux density.

If we now open a switch and remove the magnetizing current flowing through the coil we would expect the magnetic field around the coil to disappear as the magnetic flux reduced to zero. However, the magnetic flux does not completely disappear as the electromagnetic core material still retains some of its magnetism even when the current has stopped flowing in the coil.

This ability to retain some magnetism in the core after magnetization has stopped is called Retentivity or Remanence while the amount of flux density still present in the core is called Residual Magnetism, B_r.

The reason for this that some of the tiny molecular magnets do not return to a completely random pattern and still point in the direction of the original magnetising field giving them a sort of"memory". Some ferromagnetic materials have a high retentivity (magnetically hard) making them excellent for producing permanent magnets. While other ferromagnetic materials have low retentivity (magnetically soft) making them ideal for use in electromagnets, solenoids or relays. One way to reduce the this residual flux density to zero is to reverse the direction of current flow through the coil making the value of H, the magnetic field strength negative and this is called a Coersive Force.

If this reverse current is increased further the flux density will also increase in the reverse direction until the ferromagnetic core reaches saturation again but in the reverse direction from before. Reducing the magnetising current once again to zero will produce a similar amount of residual magnetism but in the reverse direction. Then by constantly changing the direction of the magnetizing current through the coil from a positive direction to a negative direction, as would be the case in an AC supply, a Magnetic Hysteresis loop of the ferromagnetic core can be produced.

MAGNETIC HYSTERESIS LOOP

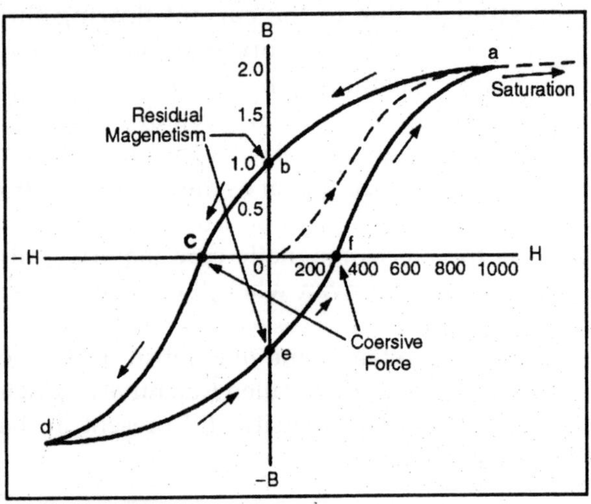

The Magnetic Hysteresis loop above, shows the behaviour of a ferromagnetic core graphically as the relationship between B and H is non-linear. Starting with an unmagnified core both B and H will be at zero, point 0 on the magnetization curve. If the magnetization current is increased in a positive direction to some value, I the magnetic field strength H increases linearly with I and the flux density B will also increase as shown by the curve from point 0 to point a as it heads towards saturation.

Now if the magnetizing current in the coil is reduced to zero the magnetic field around the core reduces to zero but the magnetic flux does not reach zero due to the residual magnetism present within the core and this is shown on the curve from point a to point b.

To reduce the flux density at point b to zero we need to reverse the current flowing through the coil. The magnetizing force which must be applied to null the the residual flux density is called a Coersive Force. This coercive force reverses the magnetic field re-arranging the molecular magnets until the core becomes unmagnified at point c.

An increase in the reverse current causes the core to be magnetized in the opposite direction and increasing this magnetization current will cause the core to reach saturation but in the opposite direction, point d on the cure which is symmetrical to point b. If the magnetizing current is reduced again to zero the residual magnetism present in the core will be equal to the previous value but in reverse at point e.

Again reversing the magnetizing current flowing through the coil this time into a positive direction will cause the magnetic flux to reach zero, point f on the curve and as before increasing the magnetization current further in a positive direction will cause the core to reach saturation at point a. Then the B-H curve follows the path of a-b-c-d-e-f-a as the magnetizing current flowing through the coil alternates between a positive and negative value such as the cycle of an AC voltage. This path is called a Magnetic Hysteresis Loop.

The effect of magnetic hysteresis shows that the magnetization process of a ferromagnetic core and therefore the flux density depends on which part of the curve the ferromagnetic core is magnetized on as this depends upon the circuits past history giving the core a form of memory. Then ferromagnetic materials have memory because they remain magnetized after the external magnetic field has been removed.

However, soft ferromagnetic materials such as iron or silicon steel have very narrow magnetic hysteresis loops resulting in very small amounts of residual magnetism making them ideal for use in relays and solenoids as they can be easily magnetized and demagnetized.

MAGNETIC HYSTERESIS LOOPS FOR SOFT AND HARD MATERIALS

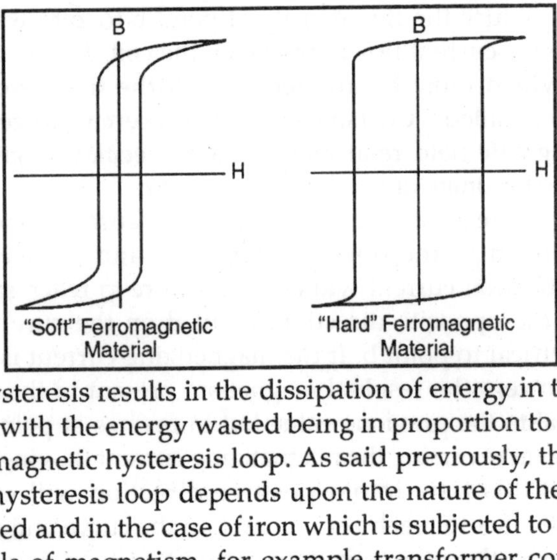

Hysteresis results in the dissipation of energy in the form of heat with the energy wasted being in proportion to the area of the magnetic hysteresis loop. As said previously, the shape of the hysteresis loop depends upon the nature of the iron or steel used and in the case of iron which is subjected to massive reversals of magnetism, for example transformer cores, it is important that the B-H hysteresis loop is as small as possible.

HYSTERESIS AND EDDY CURRENT LOSSES

HYSTERESIS

When a ferromagnetic material is magnetized in one direction, it will not relax back to zero magnetization when the imposed magnetizing field is removed. It must be driven back to zero by a field in the opposite direction. If an alternating magnetic field is applied to the material, its magnetization will trace out a loop called a hysteresis loop.

The lack of retraceability of the magnetization curve is the property called hysteresis and it is related to the existence of magnetic domains in the material. Once the magnetic domains are reoriented, it takes some energy to turn them back again. This property of ferromagnetic materials is useful as a magnetic"memory". Some compositions of ferromagnetic materials will retain an imposed magnetization indefinitely and are useful as"permanent magnets". The magnetic memory aspects of iron and chromium oxides make them useful in audio tape recording and for the magnetic storage of data on computer disks.

Hysteresis Loop

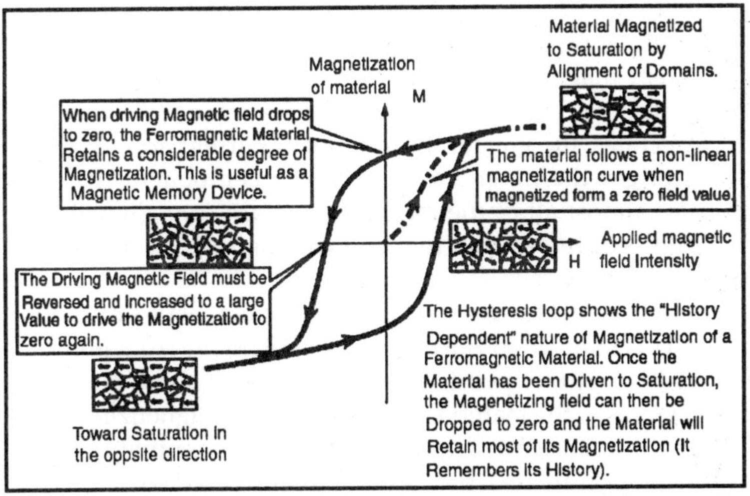

It is customary to plot the magnetization M of the sample as a function of the magnetic field strength H, since H is a measure of the externally applied field which drives the magnetization. A great deal of information can be learned about the magnetic properties of a material by studying its hysteresis loop. A hysteresis loop shows the relationship between the induced magnetic flux density (B) and the magnetizing force (H). It is often referred to as the B-H loop. An example hysteresis loop is shown below.

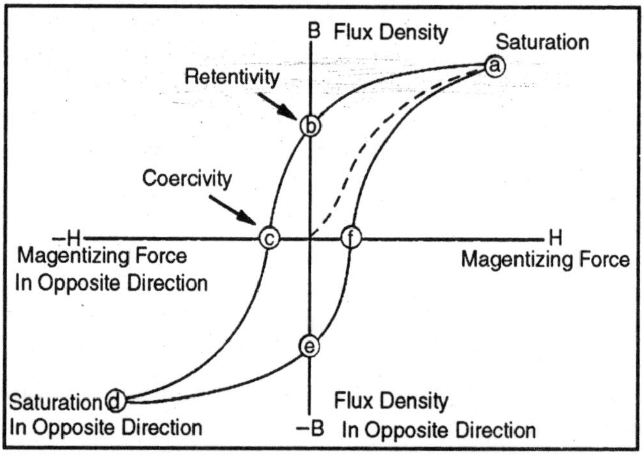

The loop is generated by measuring the magnetic flux of a ferromagnetic material while the magnetizing force is changed. A ferromagnetic material that has never been previously magnetized or has been thoroughly demagnetized will follow the dashed line as H is increased. As the line demonstrates, the greater the amount of current applied (H+), the stronger the magnetic field in the component (B+). At point"a" almost all of the magnetic domains are aligned and an additional increase in the magnetizing force will produce very little increase in magnetic flux.

The material has reached the point of magnetic saturation. When H is reduced to zero, the curve will move from point"a" to point"b." At this point, it can be seen that some magnetic flux remains in the material even though the magnetizing force is zero. This is referred to as the point of retentivity on the graph and indicates the remanence or level of residual magnetism in the material.

As the magnetizing force is reversed, the curve moves to point"c", where the flux has been reduced to zero. This is called the point of coercivity on the curve. The force required to remove the residual magnetism from the material is called the coercive force or coercivity of the material.

As the magnetizing force is increased in the negative direction, the material will again become magnetically saturated but in the opposite direction (point"d"). Reducing H to zero brings the curve to point"e." It will have a level of residual magnetism equal to that achieved in the other direction. Increasing H back in the positive direction will return B to zero. Notice that the curve did not return to the origin of the graph because some force is required to remove the residual magnetism. The curve will take a different path from point"f" back to the saturation point where it with complete the loop.

From the hysteresis loop, a number of primary magnetic properties of a material can be determined:

- *Retentivity*: A measure of the residual flux density corresponding to the saturation induction of a magnetic material. In other words, it is a material's ability to retain a certain amount of residual magnetic field when the magnetizing force is removed after achieving saturation.
- *Residual Magnetism or Residual Flux*: The magnetic flux density that remains in a material when the magnetizing force is zero. Note that residual magnetism and retentivity are the same when the material has been magnetized to the saturation point. However, the level of residual magnetism may be lower than the retentivity value when the magnetizing force did not reach the saturation level.
- *Coercive Force*: The amount of reverse magnetic field which must be applied to a magnetic material to make the magnetic flux return to zero.
- *Permeability,*: A property of a material that describes the ease with which a magnetic flux is established in the component.
- *Reluctance*: Is the opposition that a ferromagnetic material shows to the establishment of a magnetic field. Reluctance is analogous to the resistance in an electrical circuit.

Hysteresis in Magnetic Recording

Because of hysteresis, an input signal at the level indicated by the dashed line could give a magnetization anywhere between C and D, depending upon the immediate previous history of the tape. This clearly unacceptable situation is remedied by the bias signal which cycles the oxide grains around their hysteresis loops so quickly that the magnetization averages to zero when no signal is applied. The result of the bias signal is like a magnetic eddy which settles down to zero if there is no signal superimposed upon it. If there is a signal, it offsets the bias signal so that it leaves a remnant magnetization proportional to the signal offset.

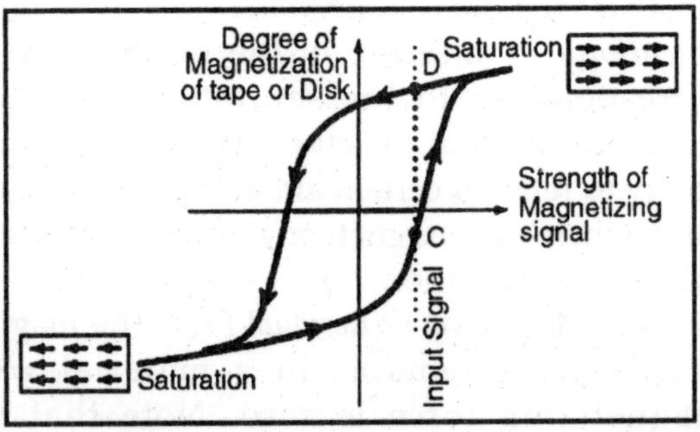

Variations in Hysteresis Curves

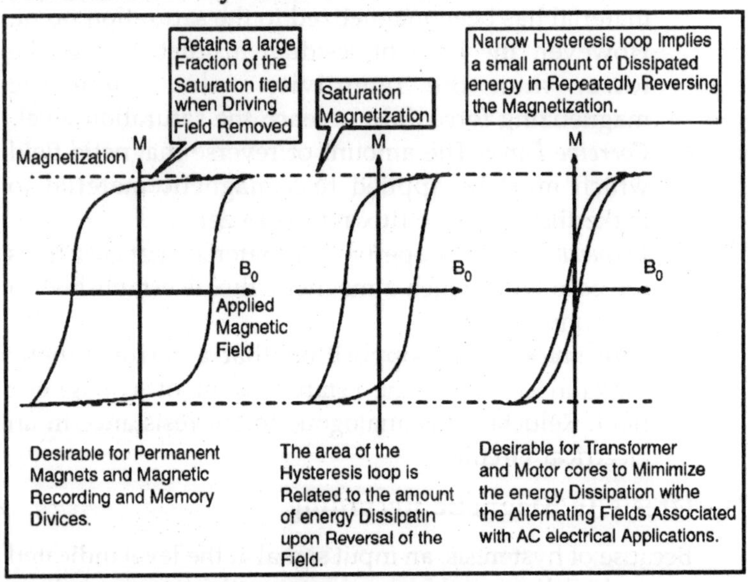

Fig. There is Considerable Variation in the Hysteresis of Different Magnetic Materials

MAGNETIC FIELD ORIENTATION

To properly inspect a component for cracks or other defects, it is important to understand that the orientation between the magnetic lines of force and the flaw is very important. There are two general types of magnetic fields that can be established within a component.

A longitudinal magnetic field has magnetic lines of force that run parallel to the long axis of the part. Longitudinal magnetization of a component can be accomplished using the longitudinal field set up by a coil or solenoid. It can also be accomplished using permanent magnets or electromagnets.

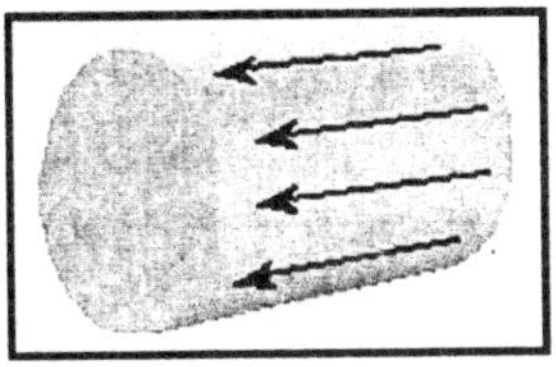

A circular magnetic field has magnetic lines of force that run circumferentially around the perimetre of a part. A circular magnetic field is induced in an article by either passing current through the component or by passing current through a conductor surrounded by the component.

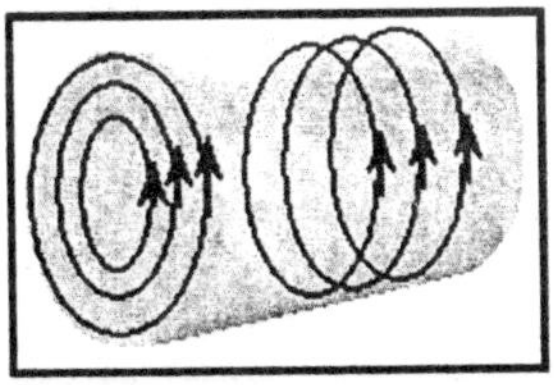

The type of magnetic field established is determined by the method used to magnetize the specimen. Being able to magnetize the part in two directions is important because the best detection of defects occurs when the lines of magnetic force are established at right angles to the longest dimension of the defect.

This orientation creates the largest disruption of the magnetic field within the part and the greatest flux leakage at the surface of the part. As can be seen in the image below, if the magnetic field is parallel to the defect, the field will see little disruption and no flux leakage field will be produced.

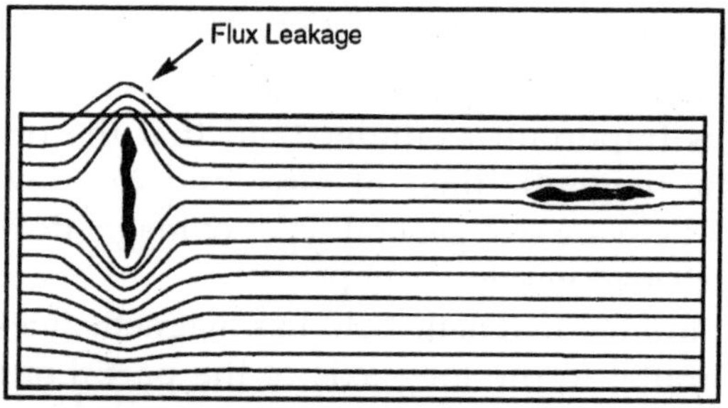

An orientation of 45 to 90 degrees between the magnetic field and the defect is necessary to form an indication. Since defects may occur in various and unknown directions, each part is normally magnetized in two directions at right angles to each other. If the component below is considered, it is known that passing current through the part from end to end will establish a circular magnetic field that will be 90 degrees to the direction of the current. Therefore, defects that have a significant dimension in the direction of the current (longitudinal defects) should be detectable. Alternately, transverse-type defects will not be detectable with circular magnetization.

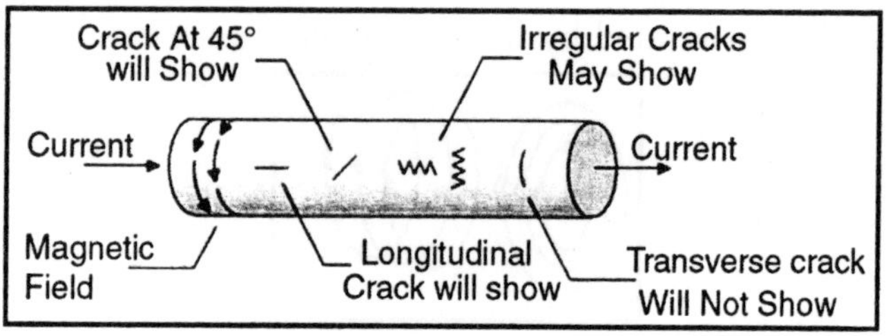

MAGNETIC CIRCUIT CALCULATIONS

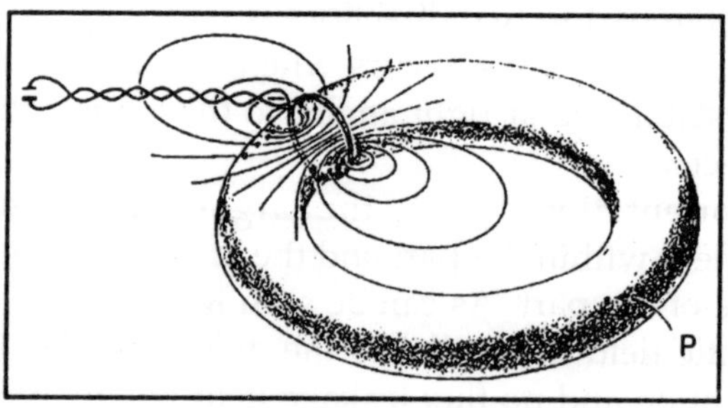

Figure above shows a ferromagnetic core around which is wound a coil of N turns carrying a current I. The magnetic flux F through some cross section of the core is to be determined.

In the absence of ferromagnetic material, the lines of B are as shown in figure above and it is of the same order of magnitude at all points within the ferromagnetic material. This is because the magnetic induction due to the current I magnetizes the core in the region near the coil and the magnetization produces Amperian currents that both increase B and extend it along the core.

This further increases and extends the magnetization and until the lines of B extend all around the core. Some of the lines of B escape into the air and then return to the core to pass again through the coil. This constitutes the leakage flux that may, or may not, be negligible. For example, if the toroid is made up of a long thin iron wire, the flux at P is negligible compared to that near the coil.

Suppose the cross section of the toroid is large enough to render the leakage flux. Then, applying Ampère's circuital law to a circular path of radius r going all around inside the toroid,

$$\oint H \cdot dl = NI,$$

$$2\pi r \cdot \frac{B}{\mu} = NI,'$$

$$B = \frac{\mu NI}{2\pi r}.$$

Then, taking R1 to be the radius corresponding to the average value of B, and R2 to be the inner radius of the toroid, the flux through the core is,

$$\Phi = \frac{\mu R_2^2}{2\pi R_1} NI.$$

This equation shows that the magnetic flux is given by the magnetomotance or another name as magneticmotive force (m.m.f.), NI multiplied by the factor,

$$\frac{\mu\pi R_2^2}{2\pi R_1},$$

this is called the permeance of the magnetic circuit. The inverse of the permeance is called as the reluctance.

Hence,

$$\text{Magnetic flux} = \text{Permeance} \times \text{Magnetomotive force}$$
$$= \text{Magnetomotive force} / \text{Reluctance}.$$

$$I = \frac{\sigma\pi R_2^2}{2\pi R_1} V.$$

Some Corresponding Quantities in Electric and Magnetic Circuit are Listed as Below.

Table.Parallel Quantities in Electric and Magnetic Circuit

Electric quantities	Magnetic quantities
Current I	Magnetic flux Φ
Current density J	Magnetic flux density B
Condudctivity σ	Permeability μ
Electromotive force	
=resistanace×I	Magnetomotive force=reluctance×Φ
Electric field intensity E	Magnetic field intensityH
Conductance =	
1/resistance	Permeance = 1/resistance
Resistance = l/σA	Reluctance = l/μA

The differences between electric and magnetic circuits are as below:

- The path of the magnetic flux flows is perpendicular to the current flows in the circuit. In other words, the directions of B and J are perpendicular.
- For a given temperature, electric resistance is constant and does not depend on current density. However, the magnetic reluctance depends on magnetic field and flux intensity since the permeability is not constant.
- Current flowing in a electric circuit involves dissipation of energy, but for magnetic circuit, energy is needed to generate magnetic flux.

MAGNETIC CIRCUIT WITH AND AIR GAP

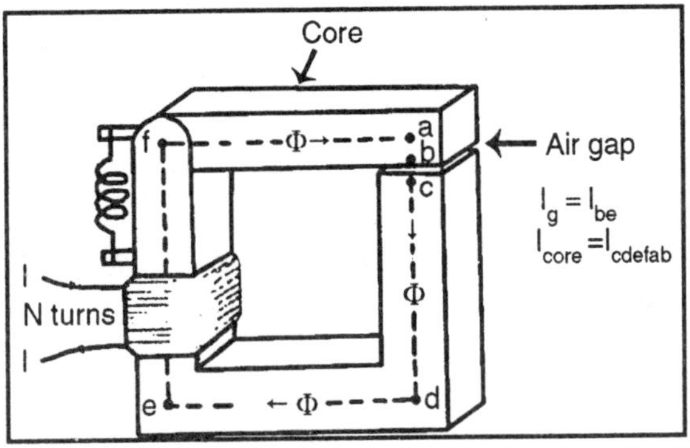

Figure above shows a simple magnetic circuit with an air gap of length lg cut in the middle of a leg. The winding provides NI ampere-turn. The spreading of the magnetic flux lines outside the common area of the core for the air gap is known as fringing field.

For simplicity, this effect is negligible and the flux distribution is assumed to be as in figure below. It can be shown that the magnetic flux generated in the air gap is equal to the magnetomotive force NI divided by the sum of the reluctances of the core and of the air gap.

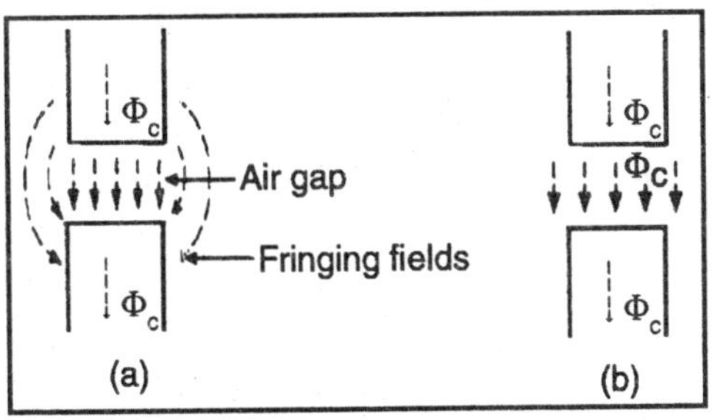

Suppose the leakage flux is negligible. Applying the Ampère's circuital law, one gets,

$$NI = H_c L_c + H_g L_g,$$

where the subscript c refers to the core and g to the air gap. The path length L_c in the core can be taken to be the length measured along the centre of the cross section of the core.

H_g can be written in terms of the magnetic flux as:

$$H_c = \frac{B_c}{\mu_c} = \frac{\Phi_c}{\mu_c A_c},$$

$$H_g = \frac{B_g}{\mu_g} = \frac{\Phi_g}{\mu_0 A_g}.$$

According to Gauss's law of magnetism, the net outward flux of B through any closed surface must be equal to zero. Hence, the flux of B must be the same over any cross section of the magnetic circuit and,

$$\oint_s B \cdot ds = 0,$$

$$\Phi_c = \Phi_g = \Phi,$$

$$B_c A_c = B_g A_g,$$

where A_c and A_g are the cross sections of the core and of the air gap, respectively. Combining above two equations gives,

$$B_s A_g \left[\frac{L_c}{\mu_c A_c} + \frac{L_g}{\mu_0 A_g} \right] = NI,$$

and the magnetic flux is,

$$\Phi = B_g A_g = \frac{NI}{\dfrac{L_c}{\mu_c A_c} + \dfrac{L_g}{\mu_0 A_g}}.$$

The denominator of above equation gives the reluctances of the core and air gap in series. Since the leakage flux is neglected, this equation gives the upper limit for Φ.

Chapter 5

Single Phase Transformer

TRANSFORMER

A transformer is a device that transfers electrical energy from one circuit to another through inductively coupled conductors-the transformer's coils.

A varying current in the first or primary winding creates a varying magnetic flux in the transformer's core, and thus a varying magnetic field through the secondary winding. This varying magnetic field induces a varying electromotive force (EMF) or"voltage" in the secondary winding. This effect is called mutual induction.

If a load is connected to the secondary, an electric current will flow in the secondary winding and electrical energy will be transferred from the primary circuit through the transformer to the load. In an ideal transformer, the induced voltage in the secondary winding (V_S) is in proportion to the primary voltage (V_P), and is given by the ratio of the number of turns in the secondary (N_S) to the number of turns in the primary (N_P) as follows:

$$\frac{V_S}{V_P} = \frac{N_S}{N_P}$$

By appropriate selection of the ratio of turns, a transformer thus allows an alternating current (AC) voltage to be"stepped up" by making NS greater than N_P, or"stepped down" by making N_S less than N_P.

In the vast majority of transformers, the windings are coils wound around a ferromagnetic core, air-core transformers being a notable exception.

Transformers range in size from a thumbnail-sized coupling transformer hidden inside a stage microphone to huge units weighing hundreds of tons used to interconnect portions of power grids.

All operate with the same basic principles, although the range of designs is wide. While new technologies have eliminated the need for transformers in some electronic circuits, transformers are still found in nearly all electronic devices designed for household ("mains") voltage. Transformers are essential for high voltage power transmission, which makes long distance transmission economically practical.

DISCOVERY

Faraday's experiment with induction between coils of wire The phenomenon of electromagnetic induction was discovered independently by Michael Faraday and Joseph Henry in 1831. However, Faraday was the first to publish the results of his experiments and thus receive credit for the discovery. The relationship between electromotive force (EMF) or"voltage" and magnetic flux was formalized in an equation now referred to as"Faraday's law of induction":

$$|\varepsilon| = \left|\frac{d\Phi_B}{dt}\right|.$$

Where:

$|\varepsilon|$ is the magnitude of the EMF in volts

Φ_B is the magnetic flux through the circuit (in webers).[3]

Induction Coils

The first type of transformer to see wide use was the induction coil, invented by Rev. Nicholas Callan of Maynooth College, Ireland in 1836.

He was one of the first researchers to realise that the more turns the secondary winding has in relation to the primary winding, the larger is the increase in EMF.

Induction coils evolved from scientists' and inventors' efforts to get higher voltages from batteries. Since batteries produce direct current (DC) rather than alternating current (AC), induction coils relied upon vibrating electrical contacts that regularly interrupted the current in the primary to create the flux changes necessary for induction. Between the 1830s and the 1870s, efforts to build better induction coils, mostly by trial and error, slowly revealed the basic principles of transformers.

In 1876, Russian engineer Pavel Yablochkov invented a lighting system based on a set of induction coils where the primary windings were connected to a source of alternating current and the secondary windings could be connected to several"electric candles" (arc lamps) of his own design. The coils Yablochkov employed functioned essentially as transformers.

Induction coils with open magnetic circuits are inefficient for transfer of power to loads. Until about 1880 the paradigm for AC power transmission from a high voltage supply to a low voltage load was a series circuit.

Open-core transformers with a ratio near 1:1 were connected with their primaries in series to allow use of a high voltage for transmission while presenting a low voltage to the lamps.

The inherent flaw in this method was that turning off a single lamp affected the voltage supplied to all others on the same circuit. Many adjustable transformer designs were introduced to compensate for this problematic characteristic of the series circuit, including those employing methods of adjusting the core or bypassing the magnetic flux around part of a coil.

The Company in Hungary began manufacturing equipment for electric lighting, and by 1883 had installed over fifty systems in Austria-Hungary. Their systems used alternating current exclusively, and included those comprising both arc and incandescent lamps, along with generators and other equipment.

Lucien Gaulard and John Dixon Gibbs first exhibited a device with an open iron core called a"secondary generator" in London in 1882, then sold the idea to the Westinghouse company in the United States. They also exhibited the invention in Turin, Italy in 1884, where it was adopted for an electric lighting system. However, the efficiency of their open-core bipolar apparatus remained low.

Efficient, practical transformer designs did not appear until the 1880s, but within a decade the transformer would be instrumental in the"War of Currents", and in seeing AC distribution systems triumph over their DC counterparts, a position in which they have remained dominant ever since.

PRINCIPLE

The transformer is based on two principles: firstly, that an electric current can produce a magnetic field (electromagnetism) and secondly that a changing magnetic field within a coil of wire induces a voltage across the ends of the coil (electromagnetic induction).

Changing the current in the primary coil changes the magnetic flux that is developed. The changing magnetic flux induces a voltage in the secondary coil. An ideal transformer is shown in the adjacent figure. Current passing through the primary coil creates a magnetic field. The primary and secondary coils are wrapped around a core of very high magnetic permeability, such as iron, so that most of the magnetic flux passes through both the primary and secondary coils.

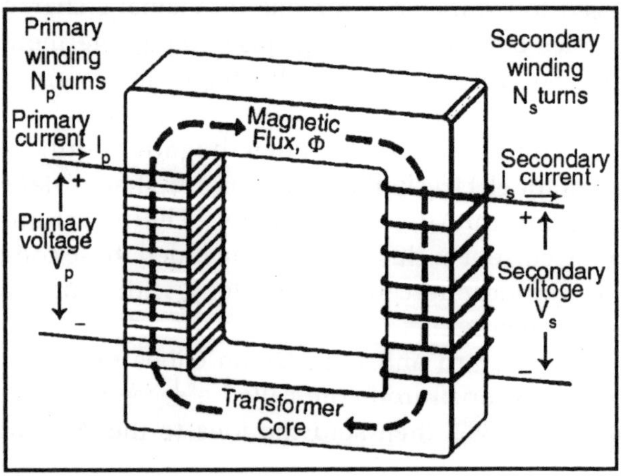

Induction Law

The voltage induced across the secondary coil may be calculated from Faraday's law of induction, which states that:

$$V_s = N_S \frac{d\Phi}{dt}$$

where V_S is the instantaneous voltage, N_S is the number of turns in the secondary coil and Φ equals the magnetic flux through one turn of the coil. If the turns of the coil are oriented perpendicular to the magnetic field lines, the flux is the product of the magnetic flux density B and the area A through which it cuts. The area is constant, being equal to the cross-sectional area of the transformer core, whereas the magnetic field varies with time according to the excitation of the primary. Since the same magnetic flux passes through both the primary and secondary coils in an ideal transformer, the instantaneous voltage across the primary winding equals:

$$V_P = N_P \frac{d\Phi}{dt}$$

Taking the ratio of the two equations for V_S and V_P gives the basic equation for stepping up or stepping down the voltage:

$$\frac{V_S}{V_P} = \frac{N_S}{N_P}$$

OPERATIONS

The simplified description above neglects several practical factors, in particular the primary current required to establish a magnetic field in the core, and the contribution to the field due to current in the secondary circuit.

Models of an ideal transformer typically assume a core of negligible reluctance with two windings of zero resistance. When a voltage is applied to the primary winding, a small current flows, driving flux around the magnetic circuit of the core. The current required to create the flux is termed the magnetizing current; since the ideal core has been assumed to have near-zero reluctance, the magnetizing current is negligible, although still required to create the magnetic field.

The changing magnetic field induces an electromotive force (EMF) across each winding. Since the ideal windings have no impedance, they have no associated voltage drop, and so the voltages V_P and V_S measured at the terminals of the transformer, are equal to the corresponding EMFs. The primary EMF, acting as it does in opposition to the primary voltage, is sometimes termed the"back EMF". This is due to Lenz's law which states that the induction of EMF would always be such that it will oppose development of any such change in magnetic field.

ENERGY LOSSES

An ideal transformer would have no energy losses, and would be 100% efficient. In practical transformers energy is dissipated in the windings, core, and surrounding structures. Larger transformers are generally more efficient, and those rated for electricity distribution usually perform better than 98%. Experimental transformers using superconducting windings achieve efficiencies of 99.85%.While the increase in efficiency is small, when applied to large heavily loaded transformers the annual savings in energy losses are significant.

A small transformer, such as a plug-in"wall wart" power adapter commonly used for low-power consumer electronics devices, may be as low as 20% efficient, with considerable energy loss even when not supplying any power to the device. Though individual losses may be only a few watts, it has been estimated that the cumulative loss from such transformers in the United States alone exceeded 32 billion kilowatt-hours (kWh) in 2002.

The losses vary with load current, and may be expressed as"no-load" or"full-load" loss. Winding resistance dominates load losses, whereas hysteresis and eddy currents losses contribute to over 99% of the no-load loss. The no-load loss can be significant, meaning that even an idle transformer constitutes a drain on an electrical supply, which encourages development of low-loss transformers (also see energy efficient transformer). Transformer losses are divided into losses in the windings, termed copper loss, and those in the magnetic circuit, termed iron loss. Losses in the transformer arise from:

Winding Resistance

Current flowing through the windings causes resistive heating of the conductors. At higher frequencies, skin effect and proximity effect create additional winding resistance and losses.

Hysteresis Losses

Each time the magnetic field is reversed, a small amount of energy is lost due to hysteresis within the core. For a given core material, the loss is proportional to the frequency, and is a function of the peak flux density to which it is subjected.

Eddy Currents

Ferromagnetic materials are also good conductors, and a solid core made from such a material also constitutes a single short-circuited turn throughout its entire length. Eddy currents therefore circulate within the core in a plane normal to the flux, and are responsible for resistive heating of the core material. The eddy current loss is a complex function of the square of supply frequency and inverse square of the material thickness.

Magnetostriction

Magnetic flux in a ferromagnetic material, such as the core, causes it to physically expand and contract slightly with each cycle of the magnetic field, an effect known as magnetostriction. This produces the buzzing sound commonly associated with transformers, and in turn causes losses due to frictional heating in susceptible cores.

Mechanical Losses

In addition to magnetostriction, the alternating magnetic field causes fluctuating electromagnetic forces between the primary and secondary windings. These incite vibrations within nearby metalwork, adding to the buzzing noise, and consuming a small amount of power.

Stray Losses

Leakage inductance is by itself largely lossless, since energy supplied to its magnetic fields is returned to the supply with the next half-cycle. However, any leakage flux that intercepts nearby conductive materials such as the transformer's support structure will give rise to eddy currents and be converted to heat. There are also radioactive losses due to the oscillating magnetic field, but these are usually small.

CIRCUIT

The physical limitations of the practical transformer may be brought together as an equivalent circuit model (shown below) built around an ideal lossless transformer. Power loss in the windings is current-dependent and is represented as in-series resistances R_P and R_S. Flux leakage results in a fraction of the applied voltage dropped without contributing to the mutual coupling, and thus can be modeled as reactances of each leakage inductance X_P and X_S in series with the perfectly coupled region.

Iron losses are caused mostly by hysteresis and eddy current effects in the core, and are proportional to the square of the core flux for operation at a given frequency. Since the core flux is proportional to the applied voltage, the iron loss can be represented by a resistance R_C in parallel with the ideal transformer.

A core with finite permeability requires a magnetizing current I_M to maintain the mutual flux in the core. The magnetizing current is in phase with the flux; saturation effects cause the relationship between the two to be non-linear, but for simplicity this effect tends to be ignored in most circuit equivalents.

With a sinusoidal supply, the core flux lags the induced EMF by 90° and this effect can be modeled as a magnetizing reactance (reactance of an effective inductance) X_M in parallel with the core loss component. R_C and X_M are sometimes together termed the magnetizing branch of the model. If the secondary winding is made open-circuit, the current I_0 taken by the magnetizing branch represents the transformer's no-load current. The secondary impedance R_S and X_S is frequently moved (or"referred") to the primary side after multiplying the components by the impedance scaling factor $(N_P/N_S)^2$.

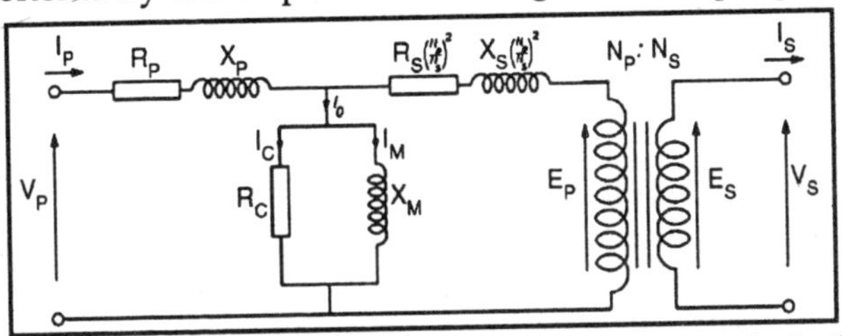

The resulting model is sometimes termed the"exact equivalent circuit", though it retains a number of approximations, such as an assumption of linearity. Analysis may be simplified by moving the magnetizing branch to the left of the primary impedance, an implicit assumption that the magnetizing current is low, and then summing primary and referred secondary impedances, resulting in so-called equivalent impedance. The parametres of equivalent circuit of a transformer can be calculated from the results of two transformer tests: open-circuit test and short-circuit test

MUTUAL INDUCTANCE

MUTUAL INDUCTANCE AND BASIC OPERATION

Suppose we were to wrap a coil of insulated wire around a loop of ferromagnetic material and energize this coil with an AC voltage source:

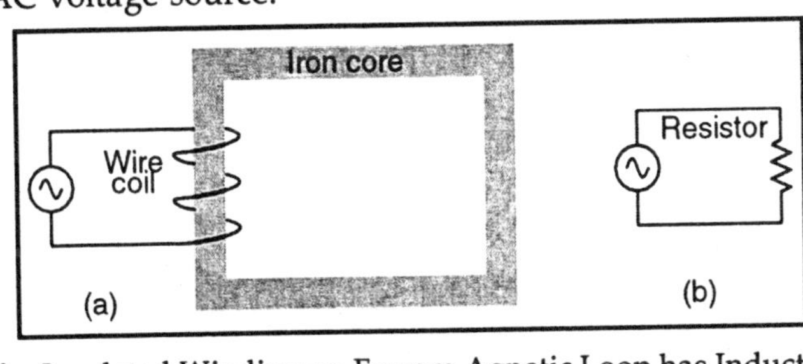

Fig. Insulated Winding on Ferrom Agnetic Loop has Inductive Reactance, Limiting AC Current.

As an inductor, we would expect this iron-core coil to oppose the applied voltage with its inductive reactance, limiting current through the coil as predicted by the equations $X_L = 2\pi fL$ and I=E/X (or I=E/Z). For the purposes of this example, though, we need to take a more detailed look at the interactions of voltage, current, and magnetic flux in the device.

Kirchhoff's voltage law describes how the algebraic sum of all voltages in a loop must equal zero. In this example, we could apply this fundamental law of electricity to describe the respective voltages of the source and of the inductor coil. Here, as in any one-source, one-load circuit, the voltage dropped across the load must equal the voltage supplied by the source, assuming zero voltage dropped along the resistance of any connecting wires.

In other words, the load (inductor coil) must produce an opposing voltage equal in magnitude to the source, in order that it may balance against the source voltage and produce an algebraic loop voltage sum of zero. From where does this opposing voltage arise?

If the load were a resistor (Figure above), the voltage drop originates from electrical energy loss, the"friction" of electrons flowing through the resistance. With a perfect inductor (no resistance in the coil wire), the opposing voltage comes from another mechanism: the reaction to a changing magnetic flux in the iron core. When AC current changes, flux Φ changes. Changing flux induces a counter EMF.

Michael Faraday discovered the mathematical relationship between magnetic flux (Φ) and induced voltage with this equation:

$$e = N\frac{d\Phi}{dt}$$

Where:

e = (Instantaneous) induced voltage in volts

N = Number of turns in wire coil (straight wire = 1)

F = Magneitc flux in Webers

t = Time in seconds

The instantaneous voltage (voltage dropped at any instant

in time) across a wire coil is equal to the number of turns of that coil around the core (N) multiplied by the instantaneous rate-of-change in magnetic flux ($d\Phi/dt$) linking with the coil. Graphed, (Figure below) this shows itself as a set of sine waves (assuming a sinusoidal voltage source), the flux wave 90° lagging behind the voltage wave:

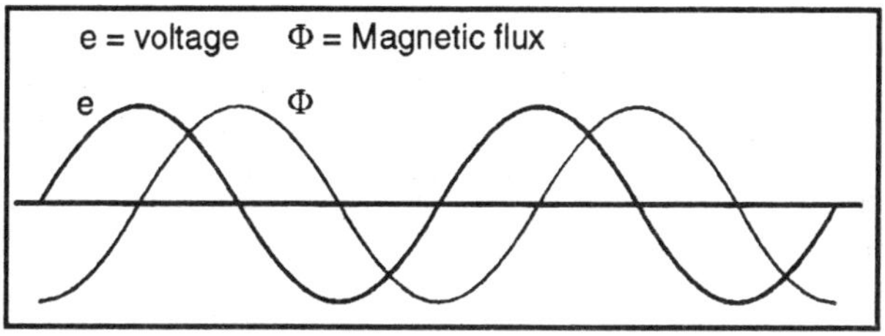

Fig. Magnetic Flux lags Applied Voltage by 90° because Flux is Proportional to a rate of change, $d\Phi/dt$.

Magnetic flux through a ferromagnetic material is analogous to current through a conductor: it must be motivated by some force in order to occur. In electric circuits, this motivating force is voltage (a.k.a. electromotive force, or EMF). In magnetic"circuits," this motivating force is magnetomotive force, or mmf. Magnetomotive force (mmf) and magnetic flux (Φ) are related to each other by a property of magnetic materials known as reluctance (the latter quantity symbolized by a strange-looking letter"R"):

Table. A comparison of "Ohm's Law" for Electric and Magnetic Circuits

$E = 1R$	$mmf = \Phi\Re$
Electrical	Magnetic

In our example, the mmf required to produce this changing magnetic flux (Φ) must be supplied by a changing current through the coil. Magnetomotive force generated by an electromagnet coil is equal to the amount of current through that coil (in amps) multiplied by the number of turns of that coil around the core (the SI unit for mmf is the amp-turn). Because the mathematical relationship between magnetic flux and mmf is directly proportional, and because the mathematical relationship between mmf and current is also directly proportional (no rates-of-change present in either equation), the current through the coil will be in-phase with the flux wave as in (Figure below)

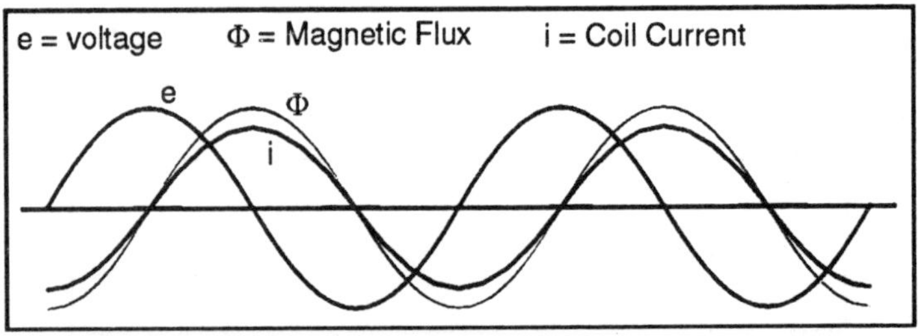

Fig. Magnetic flux, like Current, lags Applied Voltage by 90°.

This is why alternating current through an inductor lags the applied voltage waveform by 90°: because that is what is required to produce a changing magnetic flux whose rate-of-change produces an opposing voltage in-phase with the applied voltage. Due to its function in providing magnetizing force (mmf) for the core, this current is sometimes referred to as the magnetizing current.

It should be mentioned that the current through an iron-core inductor is not perfectly sinusoidal (sine-wave shaped), due to the nonlinear B/H magnetization curve of iron. In fact, if the inductor is cheaply built, using as little iron as possible, the magnetic flux density might reach high levels (approaching saturation), resulting in a magnetizing current waveform that looks something like Figure below

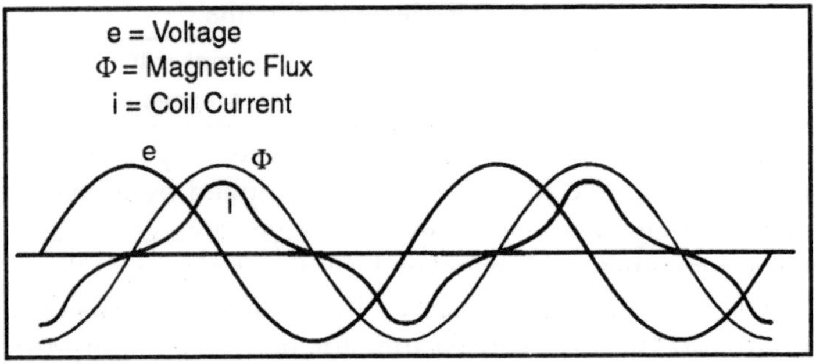

Fig. As flux Density Approaches Saturation, the Magnetizing Current Waveform Becomes Distorted.

When a ferromagnetic material approaches magnetic flux saturation, disproportionately greater levels of magnetic field force (mmf) are required to deliver equal increases in magnetic field flux (Φ).

Because mmf is proportional to current through the magnetizing coil (mmf = NI, where"N" is the number of turns of wire in the coil and"I" is the current through it), the large increases of mmf required to supply the needed increases in flux results in large increases in coil current. Thus, coil current increases dramatically at the peaks in order to maintain a flux waveform that isn't distorted, accounting for the bell-shaped half-cycles of the current waveform in the above plot.

The situation is further complicated by energy losses within the iron core. The effects of hysteresis and eddy currents conspire to further distort and complicate the current waveform, making it even less sinusoidal and altering its phase to be lagging slightly less than 90° behind the applied voltage waveform. This coil current resulting from the sum total of all magnetic effects in the core (dΦ/dt magnetization plus hysteresis losses, eddy current losses, etc.)

is called the exciting current. The distortion of an iron-core inductor's exciting current may be minimized if it is designed for and operated at very low flux densities. Generally speaking, this requires a core with large cross-sectional area, which tends to make the inductor bulky and expensive. For the sake of simplicity, though, we'll assume that our example core is far from saturation and free from all losses, resulting in a perfectly sinusoidal exciting current.

As we've seen already in the inductors chapter, having a current waveform 90° out of phase with the voltage waveform creates a condition where power is alternately absorbed and returned to the circuit by the inductor. If the inductor is perfect (no wire resistance, no magnetic core losses, etc.), it will dissipate zero power.

Let us now consider the same inductor device, except this time with a second coil (Figure below) wrapped around the same iron core. The first coil will be labeled the primary coil, while the second will be labeled the secondary:

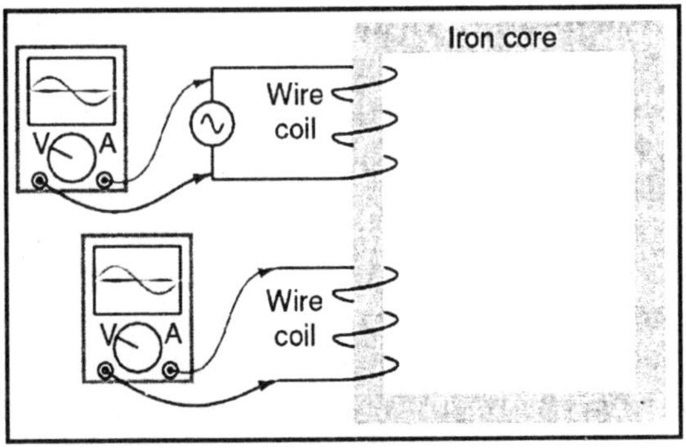

Fig. Ferromagnetic Core with Primary Coil (AC Driven) and Secondary Coil.

If this secondary coil experiences the same magnetic flux change as the primary (which it should, assuming perfect containment of the magnetic flux through the common core), and has the same number of turns around the core, a voltage of equal magnitude and phase to the applied voltage will be induced along its length. In the following graph, (Figure below) the induced voltage waveform is drawn slightly smaller than the source voltage waveform simply to distinguish one from the other:

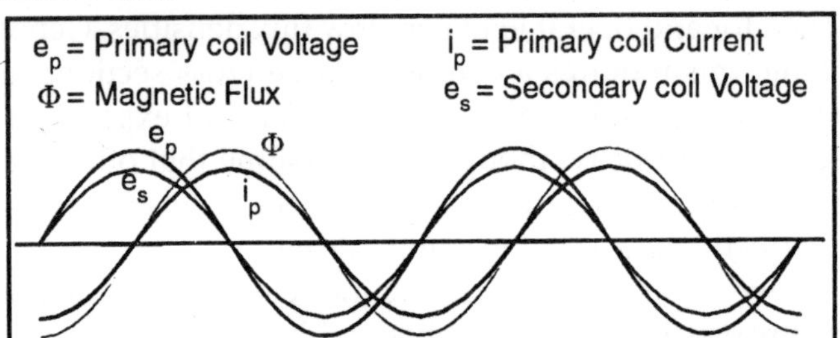

Fig. Open Circuited Secondary Sees the same Flux Φ as the Primary. Therefore Induced Secondary Voltage e_s is the same Magnitude and phase as the Primary voltage e_p.

This effect is called mutual inductance: the induction of a voltage in one coil in response to a change in current in the other coil. Like normal (self-) inductance, it is measured in the unit of Henrys, but unlike normal inductance it is symbolized by the capital letter"M" rather than the letter"L":

Inductance:

$$e = L\frac{di}{dt}$$

Mutual inductance:

$$e_2 = M\frac{di_1}{dt}$$

Where:

e_2 = voltage induced in secondary coil

i_1 = current in primary coil

No current will exist in the secondary coil, since it is open-circuited. However, if we connect a load resistor to it, an alternating current will go through the coil, in-phase with the induced voltage (because the voltage across a resistor and the current through it are always in-phase with each other).

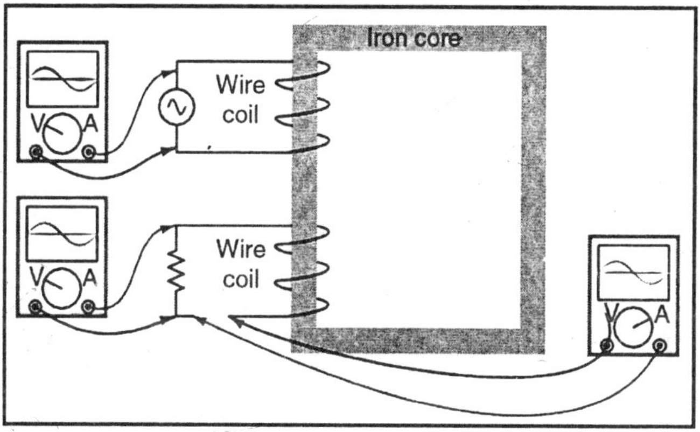

Fig. Resistive load on Secondary has Voltage and Current in-phase.

At first, one might expect this secondary coil current to cause additional magnetic flux in the core. In fact, it does not. If more flux were induced in the core, it would cause more voltage to be induced voltage in the primary coil (remember that $e = d\Phi/dt$). This cannot happen, because the primary coil's induced voltage must remain at the same magnitude and phase in order to balance with the applied voltage, in accordance with Kirchhoff's voltage law. Consequently, the magnetic flux in the core cannot be affected by secondary coil current. However, what does change is the amount of mmf in the magnetic circuit.

Magnetomotive force is produced any time electrons move

through a wire. Usually, this mmf is accompanied by magnetic flux, in accordance with the mmf=ΦR"magnetic Ohm's Law" equation.

In this case, though, additional flux is not permitted, so the only way the secondary coil's mmf may exist is if a counteracting mmf is generated by the primary coil, of equal magnitude and opposite phase.

Indeed, this is what happens, an alternating current forming in the primary coil -- 180o out of phase with the secondary coil's current -- to generate this counteracting mmf and prevent additional core flux. Polarity marks and current direction arrows have been added to the illustration to clarify phase relations:

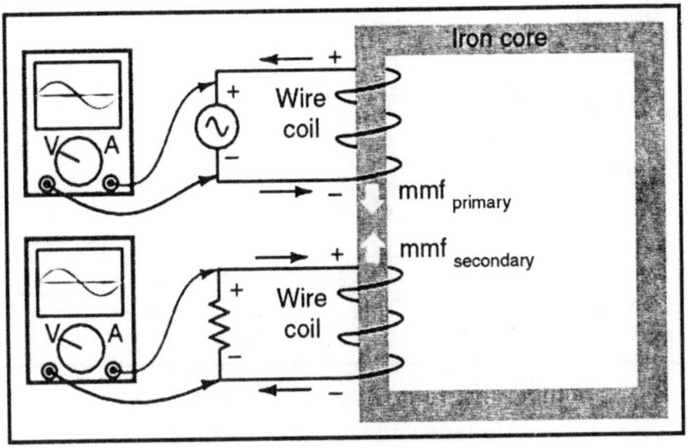

Fig. Flux remains constant with Application of a load. However, a Counteracting mmf is Produced by the loaded Secondary.

If you find this process a bit confusing, do not worry. Transformer dynamics is a complex subject. What is important to understand is this: when an AC voltage is applied to the primary coil, it creates a magnetic flux in the core, which induces AC voltage in the secondary coil in-phase with the source voltage. Any current drawn through the secondary coil to power a load induces a corresponding current in the primary coil, drawing current from the source.

Notice how the primary coil is behaving as a load with respect to the AC voltage source, and how the secondary coil is behaving as a source with respect to the resistor. Rather than energy merely being alternately absorbed and returned the primary coil circuit, energy is now being coupled to the secondary coil where it is delivered to a dissipative (energy-consuming) load.

As far as the source"knows," its directly powering the resistor. Of course, there is also an additional primary coil current lagging the applied voltage by 90°, just enough to magnetize the core to create the necessary voltage for balancing against the source (the exciting current).

We call this type of device a transformer, because it transforms electrical energy into magnetic energy, then back into electrical energy again.

Because its operation depends on electromagnetic induction between two stationary coils and a magnetic flux of changing magnitude and"polarity," transformers are necessarily AC devices. Its schematic symbol looks like two inductors (coils) sharing the same magnetic core:

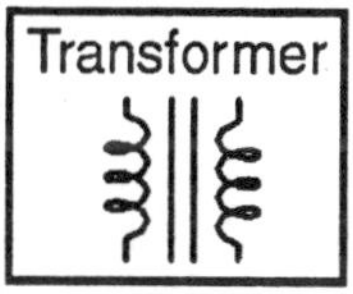

Fig. Schematic Symbol for Transformer Consists of two Inductor Symbols, Separated by lines Indicating a Ferromagnetic core.

The two inductor coils are easily distinguished in the above symbol. The pair of vertical lines represent an iron core common to both inductors. While many transformers have ferromagnetic core materials, there are some that do not, their constituent inductors being magnetically linked together through the air.

The following photograph shows a power transformer of the type used in gas-discharge lighting. Here, the two inductor coils can be clearly seen, wound around an iron core. While most transformer designs enclose the coils and core in a metal frame for protection, this particular transformer is open for viewing and so serves its illustrative purpose well:

Fig. Example of a Gas-discharge Lighting Transformer.

Both coils of wire can be seen here with copper-coloured varnish insulation. The top coil is larger than the bottom coil, having a greater number of"turns" around the core. In transformers, the inductor coils are often referred to as windings, in reference to the manufacturing process where wire is wound around the core material. As modeled in our initial example, the powered inductor of a transformer is called the primary winding, while the unpowered coil is called the secondary winding.

In the next photograph, Figure below, a transformer is shown cut in half, exposing the cross-section of the iron core as well as both windings.

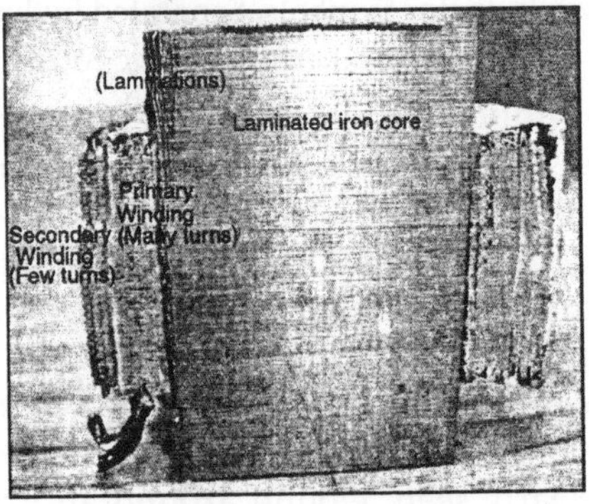

Fig. Transformer Cross-section Cut Shows Core and Windings.

Like the transformer shown previously, this unit also utilizes primary and secondary windings of differing turn counts. The wire gauge can also be seen to differ between primary and secondary windings. The reason for this disparity in wire gauge will be made clear. Additionally, the iron core can be seen in this photograph to be made of many thin sheets (laminations) rather than a solid piece. The reason for this will also be explained in this chapter.

It is easy to demonstrate simple transformer action using SPICE, setting up the primary and secondary windings of the simulated transformer as a pair of"mutual" inductors. The coefficient of magnetic field coupling is given at the end of the"k" line in the SPICE circuit description, this example being set very nearly at perfection (1.000). This coefficient describes how closely"linked" the two inductors are, magnetically. The better these two inductors are magnetically coupled, the more efficient the energy transfer between them should be.

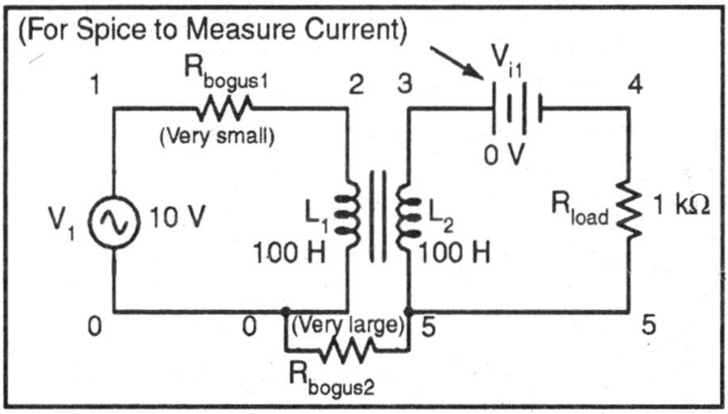

Fig. Spice Circuit for Coupled Inductors.

```
transformer
v1 1 0 ac 10 sin
rbogus1 1 2 1e-12
rbogus2 5 0 9e12
l1 2 0 100
l2 3 5 100
** This line tells SPICE that the two inductors
** l1 and l2 are magnetically"linked" together
k l1 l2 0.999
vi1 3 4 ac 0
rload 4 5 1k
.ac lin 1 60 60
```

```
.print ac v(2,0) i(v1)
.print ac v(3,5) i(vi1)
.end
```

Note: the Rbogus resistors are required to satisfy certain quirks of SPICE. The first breaks the otherwise continuous loop between the voltage source and L_1 which would not be permitted by SPICE.

The second provides a path to ground (node 0) from the secondary circuit, necessary because SPICE cannot function with any ungrounded circuits.

```
freq v(2) i(v1)
6.000E+01 1.000E+01 9.975E-03 Primary winding
freq v(3,5) i(vi1)
6.000E+01 9.962E+00 9.962E-03 Secondary winding
```

Note that with equal inductances for both windings (100 Henrys each), the AC voltages and currents are nearly equal for the two. The difference between primary and secondary currents is the magnetizing current spoken of earlier: the 90° lagging current necessary to magnetize the core.

As is seen here, it is usually very small compared to primary current induced by the load, and so the primary and secondary currents are almost equal.

What you are seeing here is quite typical of transformer efficiency. Anything less than 95% efficiency is considered poor for modern power transformer designs, and this transfer of power occurs with no moving parts or other components subject to wear.

If we decrease the load resistance so as to draw more current with the same amount of voltage, we see that the current through the primary winding increases in response.

Even though the AC power source is not directly connected to the load resistance (rather, it is electromagnetically" coupled"), the amount of current drawn from the source will be almost the same as the amount of current that would be drawn if the load were directly connected to the source.

Take a close look at the next two SPICE simulations, showing what happens with different values of load resistors:

```
transformer
v1 1 0 ac 10 sin
rbogus1 1 2 1e-12
rbogus2 5 0 9e12
l1 2 0 100
l2 3 5 100
k l1 l2 0.999
vi1 3 4 ac 0
** Note load resistance value of 200 ohms
rload 4 5 200
.ac lin 1 60 60
.print ac v(2,0) i(v1)
.print ac v(3,5) i(vi1)
.end
freq v(2) i(v1)
6.000E+01 1.000E+01 4.679E-02
freq v(3,5) i(vi1)
6.000E+01 9.348E+00 4.674E-02
```

Notice how the primary current closely follows the secondary current. In our first simulation, both currents were approximately 10 mA, but now they are both around 47 mA. In this second simulation, the two currents are closer to equality, because the magnetizing current remains the same as before while the load current has increased. Note also how the secondary voltage has decreased some with the heavier (greater current) load. Let's try another simulation with an even lower value of load resistance (15 Ω):

```
transformer
v1 1 0 ac 10 sin
rbogus1 1 2 1e-12
rbogus2 5 0 9e12
l1 2 0 100
l2 3 5 100
k l1 l2 0.999
vi1 3 4 ac 0
rload 4 5 15
.ac lin 1 60 60
.print ac v(2,0) i(v1)
```

```
.print ac v(3,5) i(vil)
.end
freq v(2) i(v1)
6.000E+01 1.000E+01 1.301E-01
freq v(3,5) i(vil)
6.000E+01 1.950E+00 1.300E-01
```

Our load current is now 0.13 amps, or 130 mA, which is substantially higher than the last time. The primary current is very close to being the same, but notice how the secondary voltage has fallen well below the primary voltage (1.95 volts versus 10 volts at the primary).

The reason for this is an imperfection in our transformer design: because the primary and secondary inductances aren't perfectly linked (a k factor of 0.999 instead of 1.000) there is"stray" or"leakage" inductance. In other words, some of the magnetic field isn't linking with the secondary coil, and thus cannot couple energy to it:

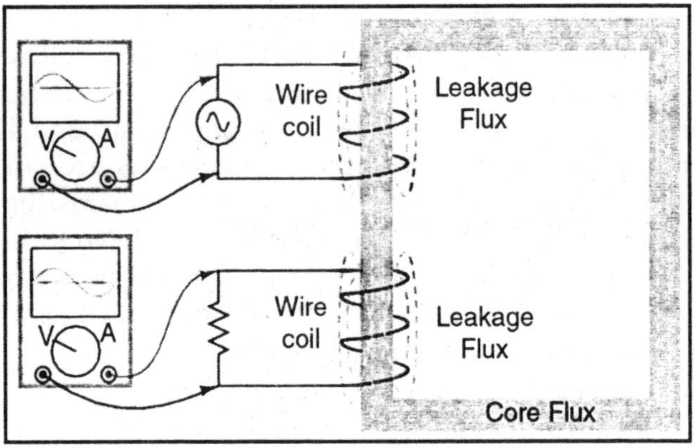

Fig. Leakage Inductance is due to Magnetic Flux not Cutting both Windings.

Consequently, this"leakage" flux merely stores and returns energy to the source circuit via self-inductance, effectively acting as a series impedance in both primary and secondary circuits. Voltage gets dropped across this series impedance, resulting in a reduced load voltage: voltage across the load"sags" as load current increases.

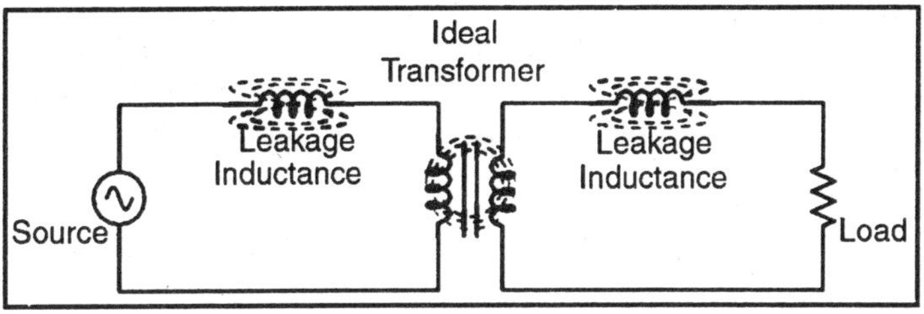

Fig. Equivalent Circuit models Leakage Inductance as series Inductors Independent of the"ideal Transformer".

If we change the transformer design to have better magnetic coupling between the primary and secondary coils, the figures for voltage between primary and secondary windings will be much closer to equality again:

```
transformer
v1 1 0 ac 10 sin
rbogus1 1 2 1e-12
rbogus2 5 0 9e12
l1 2 0 100
l2 3 5 100
** Coupling factor = 0.99999 instead of 0.999
k l1 l2 0.99999
vil 3 4 ac 0
rload 4 5 15
.ac lin 1 60 60
.print ac v(2,0) i(v1)
.print ac v(3,5) i(vil)
.end
freq v(2) i(v1)
6.000E+01 1.000E+01 6.658E-01
freq v(3,5) i(vil)
6.000E+01 9.987E+00 6.658E-01
```

Here we see that our secondary voltage is back to being equal with the primary, and the secondary current is equal to the primary current as well. Unfortunately, building a real transformer with coupling this complete is very difficult. A compromise solution is to design both primary and secondary coils with less inductance, the strategy being that less inductance overall leads to less"leakage" inductance to cause trouble, for any given degree of magnetic coupling inefficiency. This results in a load voltage that is closer to ideal with the same (high current heavy) load and the same coupling factor:

```
transformer
v1 1 0 ac 10 sin
rbogus1 1 2 1e-12
rbogus2 5 0 9e12
** inductance = 1 henry instead of 100 henrys
l1 2 0 1
l2 3 5 1
k l1 l2 0.999
vi1 3 4 ac 0
rload 4 5 15
.ac lin 1 60 60
.print ac v(2,0) i(v1)
.print ac v(3,5) i(vi1)
.end
freq v(2) i(v1)
6.000E+01 1.000E+01 6.664E-01
freq v(3,5) i(vi1)
6.000E+01 9.977E+00 6.652E-01
```

Simply by using primary and secondary coils of less inductance, the load voltage for this heavy load (high current) has been brought back up to nearly ideal levels (9.977 volts). At this point, one might ask,"If less inductance is all that's needed to achieve near-ideal performance under heavy load, then why worry about coupling efficiency at all? If its impossible to build a transformer with perfect coupling, but easy to design coils with low inductance, then why not just build all transformers with low-inductance coils and have excellent efficiency even with poor magnetic coupling?"

The answer to this question is found in another simulation: the same low-inductance transformer, but this time with a lighter load (less current) of 1 kΩ instead of 15 Ω:

```
transformer
v1 1 0 ac 10 sin
rbogus1 1 2 1e-12
```

```
rbogus2 5 0 9e12
l1 2 0 1
l2 3 5 1
k l1 l2 0.999
vil 3 4 ac 0
rload 4 5 1k
.ac lin 1 60 60
.print ac v(2,0) i(v1)
.print ac v(3,5) i(vil)
.end
freq v(2) i(v1)
6.000E+01 1.000E+01 2.835E-02
freq v(3,5) i(vil)
6.000E+01 9.990E+00 9.990E-03
```

With lower winding inductances, the primary and secondary voltages are closer to being equal, but the primary and secondary currents are not. In this particular case, the primary current is 28.35 mA while the secondary current is only 9.990 mA: almost three times as much current in the primary as the secondary. Why is this? With less inductance in the primary winding, there is less inductive reactance, and consequently a much larger magnetizing current. A substantial amount of the current through the primary winding merely works to magnetize the core rather than transfer useful energy to the secondary winding and load.

An ideal transformer with identical primary and secondary windings would manifest equal voltage and current in both sets of windings for any load condition. In a perfect world, transformers would transfer electrical power from primary to secondary as smoothly as though the load were directly connected to the primary power source, with no transformer there at all. However, you can see this ideal goal can only be met if there is perfect coupling of magnetic flux between primary and secondary windings.

Being that this is impossible to achieve, transformers must be designed to operate within certain expected ranges of voltages and loads in order to perform as close to ideal as possible. For now, the most important thing to keep in mind is a transformer's basic operating principle: the transfer of power from the primary to the secondary circuit via electromagnetic coupling.

ELECTRICAL ISOLATION

Aside from the ability to easily convert between different levels of voltage and current in AC and DC circuits, transformers also provide an extremely useful feature called isolation, which is the ability to couple one circuit to another without the use of direct wire connections. We can demonstrate an application of this effect with another SPICE simulation: this time showing"ground" connections for the two circuits, imposing a high DC voltage between one circuit and ground through the use of an additional voltage source:

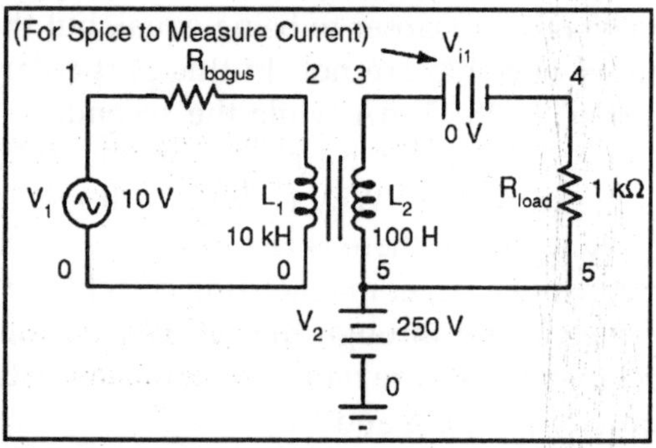

Fig.Transformer Isolates 10 V_{ac} at V_1 from 250 VDC at V_2.

```
v1 1 0 ac 10 sin
rbogus1 1 2 1e-12
v2 5 0 dc 250
l1 2 0 10000
l2 3 5 100
k l1 l2 0.999
vi1 3 4 ac 0
rload 4 5 1k
.ac lin 1 60 60
.print ac v(2,0) i(v1)
.print ac v(3,5) i(vi1)
.end
```

```
DC voltages referenced to ground (node 0):
(1) 0.0000 (2) 0.0000 (3) 250.0000
(4) 250.0000 (5) 250.0000
AC voltages:
freq v(2) i(v1)
6.000E+01 1.000E+01 9.975E-05 Primary winding
freq v(3,5) i(vi1)
6.000E+01 9.962E-01 9.962E-04 Secondary winding
```

SPICE shows the 250 volts DC being impressed upon the secondary circuit elements with respect to ground, (Figure above) but as you can see there is no effect on the primary circuit (zero DC voltage) at nodes 1 and 2, and the transformation of AC power from primary to secondary circuits remains the same as before. The impressed voltage in this example is often called a common-mode voltage because it is seen at more than one point in the circuit with reference to the common point of ground. The transformer isolates the common-mode voltage so that it is not impressed upon the primary circuit at all, but rather isolated to the secondary side. For the record, it does not matter that the common-mode voltage is DC, either. It could be AC, even at a different frequency, and the transformer would isolate it from the primary circuit all the same. There are applications where electrical isolation is needed between two AC circuit without any transformation of voltage or current levels. In these instances, transformers called isolation transformers having 1:1 transformation ratios are used ourselves:

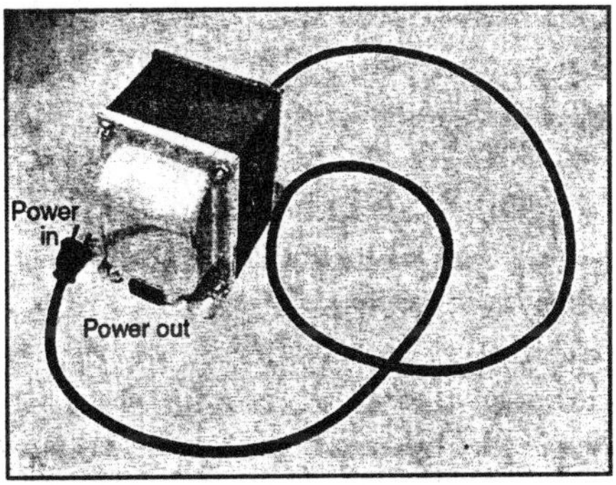

Fig. Isolation Transformer Isolates Power out from the Power line.

PHASING

Since transformers are essentially AC devices, we need to be aware of the phase relationships between the primary and secondary circuits. Using our SPICE example from before, we can plot the waveshapes for the primary and secondary circuits and see the phase relations for

```
spice transient analysis file for use with nutmeg:
transformer
v1 1 0 sin(0 15 60 0 0)
rbogus1 1 2 1e-12
v2 5 0 dc 250
l1 2 0 10000
l2 3 5 100
k l1 l2 0.999
vi1 3 4 ac 0
rload 4 5 1k
.tran 0.5m 17m
.end
nutmeg commands:
setplot tran1
plot v(2) v(3,5)
```

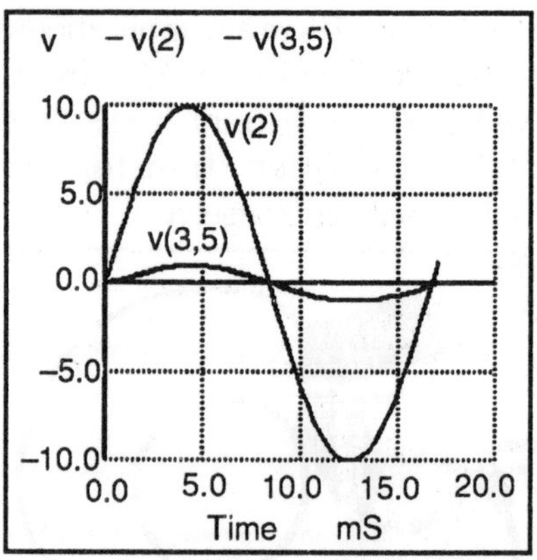

Fig. Secondary Voltage V(3,5) is in-phase with Primary Voltage V(2), and Stepped down by Factor of ten.

In going from primary, V(2), to secondary, V(3,5), the voltage was stepped down by a factor of ten, and the current was stepped up by a factor of 10. Both current and voltage waveforms are in-phase in going from primary to secondary.

```
nutmeg commands:
setplot tran1
plot I(L1#branch) I(L2#branch)
```

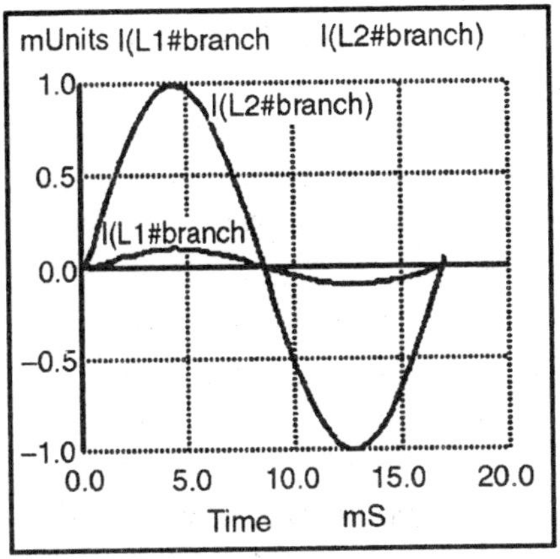

Fig. Primary and Secondary Currents are in-phase. Secondary Current is Stepped up by a Factor of ten.

It would appear that both voltage and current for the two transformer windings are in-phase with each other, at least for our resistive load. This is simple enough, but it would be nice to know which way we should connect a transformer in order to ensure the proper phase relationships be kept. After all, a transformer is nothing more than a set of magnetically-linked inductors, and inductors don't usually come with polarity markings of any kind. If we were to look at an unmarked transformer, we would have no way of knowing which way to hook it up to a circuit to get in-phase (or 180° out-of-phase) voltage and current:

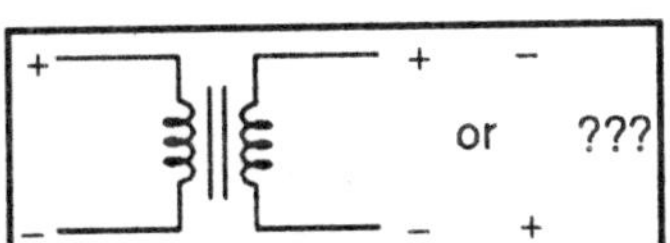

Fig. As a Practical matter, the Polarity of a Transformer can be Ambiguous.

Since this is a practical concern, transformer manufacturers have come up with a sort of polarity marking standard to denote phase relationships. It is called the dot convention, and is nothing more than a dot placed next to each corresponding leg of a transformer winding: (Figure below)

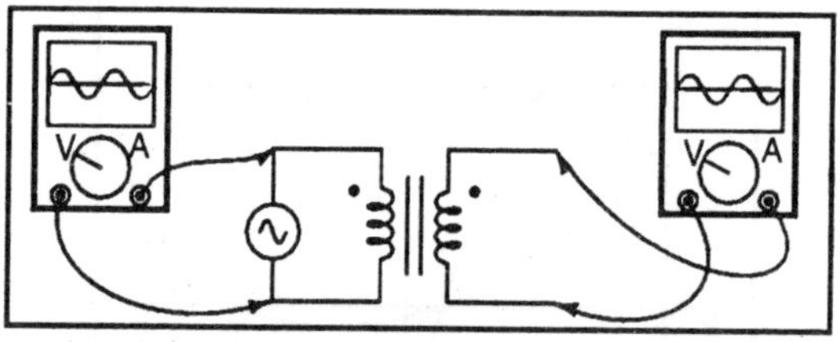

Fig.A Pair of dots Indicates like Polarity.

Typically, the transformer will come with some kind of schematic diagram labeling the wire leads for primary and secondary windings. On the diagram will be a pair of dots similar to what is seen above. Sometimes dots will be omitted, but when"H" and"X" labels are used to label transformer winding wires, the subscript numbers are supposed to represent winding polarity. The"1" wires (H_1 and X_1) represent where the polarity-marking dots would normally be placed.

The similar placement of these dots next to the top ends of the primary and secondary windings tells us that whatever instantaneous voltage polarity seen across the primary winding will be the same as that across the secondary winding. In other words, the phase shift from primary to secondary will be zero degrees.

On the other hand, if the dots on each winding of the transformer do not match up, the phase shift will be 180o between primary and secondary, like this: (Figure below)

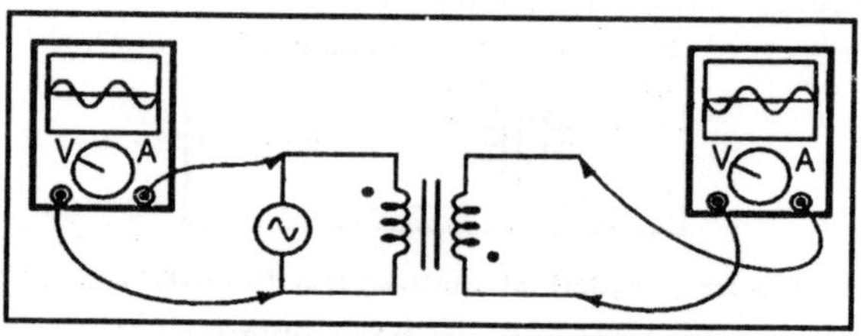

Fig.*Out of Phase*: Primary red to dot, Secondary black to dot.

Of course, the dot convention only tells you which end of each winding is which, relative to the other winding(s). If you want to reverse the phase relationship yourself, all you have to do is swap the winding connections like this:

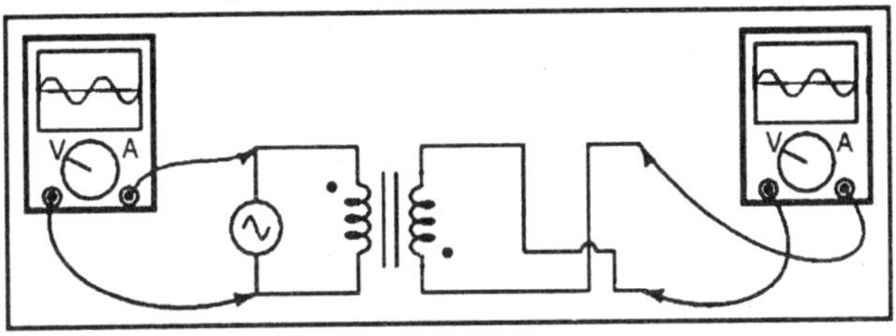

Fig. In phase: Primary red to dot, Secondary red to dot.

VOLTAGE REGULATION

As we saw in a few SPICE analyses earlier in this chapter, the output voltage of a transformer varies some with varying load resistances, even with a constant voltage input. The degree of variance is affected by the primary and secondary winding inductances, among other factors, not the least of which includes winding resistance and the degree of mutual inductance (magnetic coupling) between the primary and secondary windings. For power transformer applications, where the transformer is seen by the load (ideally) as a constant source of voltage, it is good to have the secondary voltage vary as little as possible for wide variances in load current.

The measure of how well a power transformer maintains constant secondary voltage over a range of load currents is called the transformer's voltage regulation.

It can be calculated from the following formula:

$$\text{Regulation percentage} = \frac{E_{\text{no-load}} - E_{\text{full-load}}}{E_{\text{full-load}}}(100\%)$$

"Full-load" means the point at which the transformer is operating at maximum permissible secondary current. This operating point will be determined primarily by the winding wire size (ampacity) and the method of transformer cooling. Taking our first SPICE transformer simulation as an example, let's compare the output voltage with a 1 kΩ load versus a

200 Ω load (assuming that the 200 Ω load will be our"full load" condition). *Recall if you will that our constant primary voltage was 10.00 volts AC:*

```
freq v(3,5) i(vil)
6.000E+01 9.962E+00 9.962E-03 Output with 1k ohm
load
freq v(3,5) i(vil)
6.000E+01 9.348E+00 4.674E-02 Output with 200
ohm load
```

Notice how the output voltage decreases as the load gets heavier (more current). Now let's take that same transformer circuit and place a load resistance of extremely high magnitude across the secondary winding to simulate a"no-load" condition:

```
transformer
v1 1 0 ac 10 sin
rbogus1 1 2 1e-12
rbogus2 5 0 9e12
l1 2 0 100
l2 3 5 100
k l1 l2 0.999
vil 3 4 ac 0
rload 4 5 9e12
.ac lin 1 60 60
.print ac v(2,0) i(v1)
.print ac v(3,5) i(vil)
.end
freq v(2) i(v1)
6.000E+01 1.000E+01 2.653E-04
freq v(3,5) i(vil)
6.000E+01 9.990E+00 1.110E-12 Output with (almost)
no load
```

So, we see that our output (secondary) voltage spans a range of 9.990 volts at (virtually) no load and 9.348 volts at the point we decided to call"full load." Calculating voltage regulation with these figures, we get:

$$\text{Regulation percentage} = \frac{9.990\text{V} - 9.348\text{V}}{9.348\text{V}}(100\%)$$

$$\text{Regulation percentage} = 6.8678\%$$

Incidentally, this would be considered rather poor (or"loose") regulation for a power transformer. Powering a simple resistive load like this, a good power transformer should exhibit a regulation per centage of less than 3%. Inductive loads tend to create a condition of worse voltage regulation, so this analysis with purely resistive loads was a"best-case" condition.

There are some applications, however, where poor regulation is actually desired. One such case is in discharge lighting, where a step-up transformer is required to initially generate a high voltage (necessary to"ignite" the lamps), then the voltage is expected to drop off once the lamp begins to draw current. This is because discharge lamps' voltage requirements tend to be much lower after a current has been established through the arc path. In this case, a step-up transformer with poor voltage regulation suffices nicely for the task of conditioning power to the lamp.

Another application is in current control for AC arc welders, which are nothing more than step-down transformers supplying low-voltage, high-current power for the welding process. A high voltage is desired to assist in"striking" the arc (getting it started), but like the discharge lamp, an arc doesn't require as much voltage to sustain itself once the air has been heated to the point of ionization.

Thus, a decrease of secondary voltage under high load current would be a good thing. Some arc welder designs provide arc current adjustment by means of a movable iron core in the transformer, cranked in or out of the winding assembly by the operator. Moving the iron slug away from the windings reduces the strength of magnetic coupling between the windings, which diminishes no-load secondary voltage and makes for poorer voltage regulation.

No exposition on transformer regulation could be called complete without mention of an unusual device called a ferroresonant transformer."Ferroresonance" is a phenomenon associated with the behaviour of iron cores while operating near a point of magnetic saturation (where the core is so strongly magnetized that further increases in winding current results in little or no increase in magnetic flux). While being somewhat difficult to describe without going deep into electromagnetic theory, the ferroresonant transformer is a power transformer engineered to operate in a condition of persistent core saturation. That is, its iron core is"stuffed full" of magnetic lines of flux for a large portion of the AC cycle so that variations in supply voltage (primary winding current) have little effect on the core's magnetic flux density, which means the secondary winding outputs a nearly constant voltage despite significant variations in supply (primary winding) voltage.

Normally, core saturation in a transformer results in distortion of the sinewave shape, and the ferroresonant transformer is no exception.

To combat this side effect, ferroresonant transformers have an auxiliary secondary winding paralleled with one or more capacitors, forming a resonant circuit tuned to the power supply frequency.

This"tank circuit" serves as a filter to reject harmonics created by the core saturation, and provides the added benefit of storing energy in the form of AC oscillations, which is available for sustaining output winding voltage for brief periods of input voltage loss (milliseconds' worth of time, but certainly better than nothing).

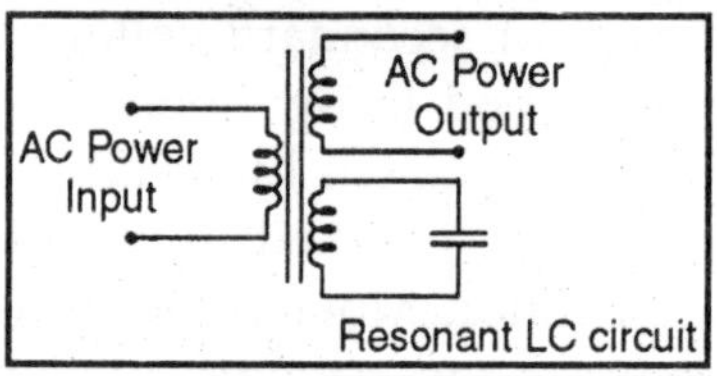

Fig. Ferroresonant Transformer Provides Voltage Regulation of the Output.

In addition to blocking harmonics created by the saturated core, this resonant circuit also"filters out" harmonic frequencies generated by nonlinear (switching) loads in the secondary winding circuit and any harmonics present in the source voltage, providing"clean" power to the load.

Ferroresonant transformers offer several features useful in AC power conditioning: constant output voltage given substantial variations in input voltage, harmonic filtering between the power source and the load, and the ability to"ride through" brief losses in power by keeping a reserve of energy in its resonant tank circuit. These transformers are also highly tolerant of excessive loading and transient (momentary) voltage surges. They are so tolerant, in fact, that some may be briefly paralleled with unsynchronized AC power sources, allowing a load to be switched from one source of power to another in a"make-before-break" fashion with no interruption of power on the secondary side!

Unfortunately, these devices have equally noteworthy disadvantages: they waste a lot of energy (due to hysteresis losses in the saturated core), generating significant heat in the process, and are intolerant of frequency variations, which means they don't work very well when powered by small engine-driven generators having poor speed regulation.

Voltages produced in the resonant winding/capacitor circuit tend to be very high, necessitating expensive capacitors and presenting the service technician with very dangerous working voltages. Some applications, though, may prioritize the ferroresonant transformer's advantages over its disadvantages. Semiconductor circuits exist to"condition" AC power as an alternative to ferroresonant devices, but none can compete with this transformer in terms of sheer simplicity.

TYPES

Potential Transformers

Transformers can also be used in electrical instrumentation systems. Due to transformers' ability to step up or step down voltage and current, and the electrical isolation they provide, they can serve as a way of connecting electrical instrumentation to high-voltage, high current power systems. Suppose we wanted to accurately measure the voltage of a 13.8 kV power system (a very common power distribution voltage in American industry):

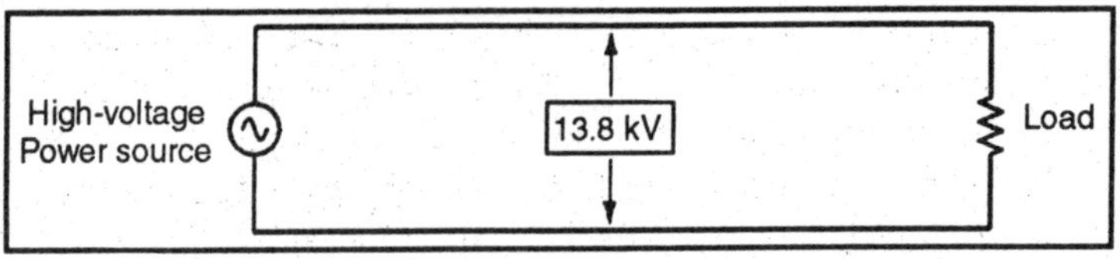

Fig. Direct Measurement of high Voltage by a Voltmetre is a Potential Safety Hazard.

Designing, installing, and maintaining a voltmetre capable of directly measuring 13,800 volts AC would be no easy task. The safety hazard alone of bringing 13.8 kV conductors into an instrument panel would be severe, not to mention the design of the voltmetre itself. However, by using a precision step-down transformer, we can reduce the 13.8 kV down to a safe level of voltage at a constant ratio, and isolate it from the instrument connections, adding an additional level of safety to the metreing system:

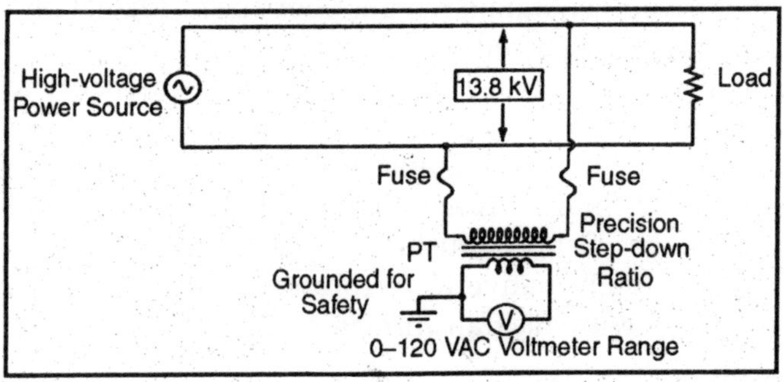

Fig. *Instrumentation Application*:"Potential Transformer" Precisely Scales Dangerous High Voltage to a Safe Value Applicable to a Conventional Voltmetre.

Now the voltmetre reads a precise fraction, or ratio, of the actual system voltage, its scale set to read as though it were measuring the voltage directly. The transformer keeps the instrument voltage at a safe level and electrically isolates it from the power system, so there is no direct connection between the power lines and the instrument or instrument wiring. When used in this capacity, the transformer is called a Potential Transformer, or simply PT.

Potential transformers are designed to provide as accurate a voltage step-down ratio as possible. To aid in precise voltage regulation, loading is kept to a minimum: the voltmetre is made to have high input impedance so as to draw as little current from the PT as possible.

As you can see, a fuse has been connected in series with the PTs primary winding, for safety and ease of disconnecting the PT from the circuit. A standard secondary voltage for a PT is 120 volts AC, for full-rated power line voltage. The standard voltmetre range to accompany a PT is 150 volts, full-scale. PTs with custom winding ratios can be manufactured to suit any application. This lends itself well to industry standardization of the actual voltmetre instruments themselves, since the PT will be sized to step the system voltage down to this standard instrument level.

Current Transformers

Following the same line of thinking, we can use a transformer to step down current through a power line so that we are able to safely and easily measure high system currents with inexpensive ammetres. Of course, such a transformer would be connected in series with the power line, like.

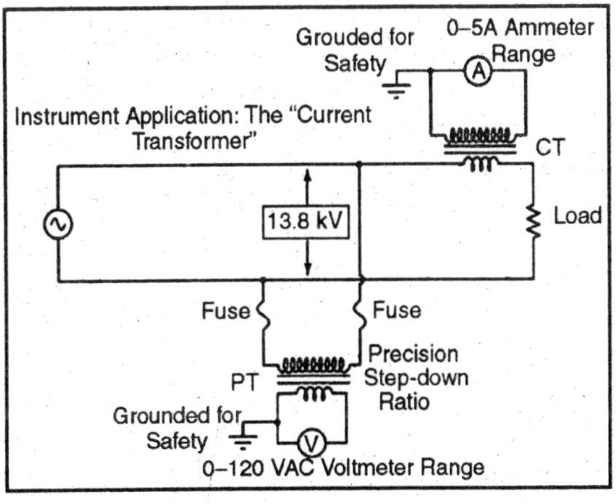

Fig. *Instrumentation Application*:"Currrent Transformer" Steps High Current down to a Value Applicable to a Conventional Ammeter.

Note that while the PT is a step-down device, the Current Transformer (or CT) is a step-up device (with respect to voltage), which is what is needed to step down the power line current. Quite often, CTs are built as donut-shaped devices through which the power line conductor is run, the power line itself acting as a single-turn primary winding:

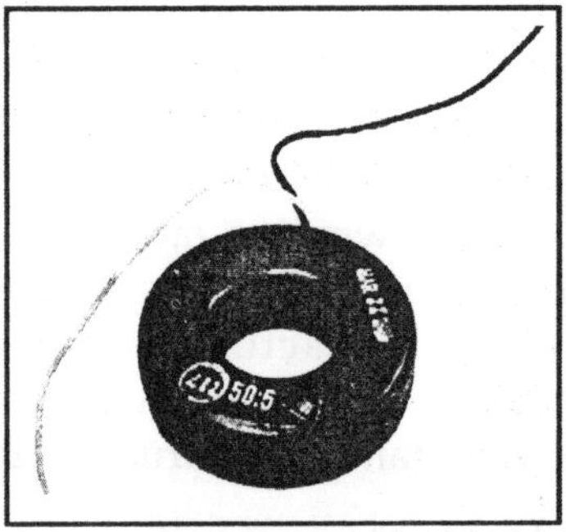

Fig. Current Conductor to be Measured is Threaded Through the Opening. Scaled down Current is Available on Wire Leads.

Some CTs are made to hinge open, allowing insertion around a power conductor without disturbing the conductor at all. The industry standard secondary current for a CT is a range of 0 to 5 amps AC. Like PTs, CTs can be made with custom winding ratios to fit almost any application. Because their"full load" secondary current is 5 amps, CT ratios are usually described in terms of full-load primary amps to 5 amps, like this:

600: 5 ratio (for measuring up to 600 A line current)

100: 5 ratio (for measuring up to 100 A line current)

lk: 5 (for measuring up to 1000 line current)

The"donut" CT shown in the photograph has a ratio of 50:5. That is, when the conductor through the centre of the torus is carrying 50 amps of current (AC), there will be 5 amps of current induced in the CT's winding.

Because CTs are designed to be powering ammeters, which are low-impedance loads, and they are wound as voltage step-up transformers, they should never, ever be operated with an open-circuited secondary winding. Failure to heed this warning will result in the CT producing extremely high secondary voltages, dangerous to equipment and personnel alike.

To facilitate maintenance of ammeter instrumentation, short-circuiting switches are often installed in parallel with the CT's secondary winding, to be closed whenever the ammeter is removed for service:

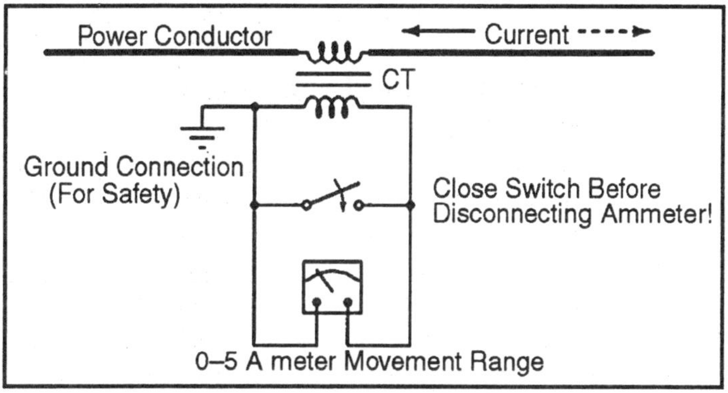

Fig. Short-circuit Switch Allows Ammeter to be Removed from an Active Current Transformer Circuit.

Though it may seem strange to intentionally short-circuit a power system component, it is perfectly proper and quite necessary when working with current transformers.

Air Core Transformers

Another kind of special transformer, seen often in radio-frequency circuits, is the air core transformer. True to its name, an air core transformer has its windings wrapped around a nonmagnetic form, usually a hollow tube of some material.

The degree of coupling (mutual inductance) between windings in such a transformer is many times less than that of an equivalent iron-core transformer, but the undesirable characteristics of a ferromagnetic core (eddy current losses, hysteresis, saturation, etc.) are completely eliminated. It is in high-frequency applications that these effects of iron cores are most problematic.

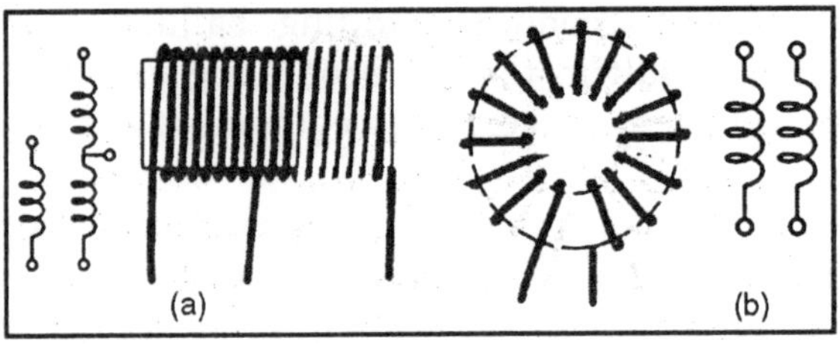

Fig. Air core Transformers may be wound on Cylindrical (a) or Toroidal (b) forms. Centre tapped Primary with Secondary (a). Bifilar WEinding on Toroidal form (b).

The inside tapped solenoid winding, without the over winding, could match unequal impedances when DC isolation is not required. When isolation is required the over winding is added over one end of the main winding. Air core transformers are used at radio frequencies when iron core losses are too high.

Frequently air core transformers are paralleled with a capacitor to tune it to resonance. The over winding is connected between a radio antenna and ground for one such application. The secondary is tuned to resonance with a variable capacitor.

The output may be taken from the tap point for amplification or detection. Small millimetre size air core transformers are used in radio receivers. The largest radio transmitters may use metre sized coils. Unshielded air core solenoid transformers are mounted at right angles to each other to prevent stray coupling.

Stray coupling is minimized when the transformer is wound on a toroid form.Toroidal air core transformers also show a higher degree of coupling, particularly for bifilar windings. Bifilar windings are wound from a slightly twisted pair of wires. This implies a 1:1 turns ratio. Three or four wires may be grouped for 1:2 and other integral ratios. Windings do not have to be bifilar. This allows arbitrary turns ratios. However, the degree of coupling suffers. Toroidal air core transformers are rare except for VHF (Very High Frequency) work. Core materials other than air such as powdered iron or ferrite are preferred for lower radio frequencies.

PRACTICAL CONSIDERATIONS

POWER CAPACITY

As has already been observed, transformers must be well designed in order to achieve acceptable power coupling, tight voltage regulation, and low exciting current distortion. Also, transformers must be designed to carry the expected values of primary and secondary winding current without any trouble. This means the winding conductors must be made of the proper gauge wire to avoid any heating problems. An ideal transformer would have perfect coupling (no leakage inductance), perfect voltage regulation, perfectly sinusoidal exciting current, no hysteresis or eddy current losses, and wire thick enough to handle any amount of current. Unfortunately, the ideal transformer would have to be infinitely large and heavy to meet these design goals. Thus, in the business of practical transformer design, compromises must be made.

Additionally, winding conductor insulation is a concern where high voltages are encountered, as they often are in step-up and step-down power distribution transformers. Not only do the windings have to be well insulated from the iron core, but each winding has to be sufficiently insulated from the other in order to maintain electrical isolation between windings. Respecting these limitations, transformers are rated for certain levels of primary and secondary winding voltage and current, though the current rating is usually derived from a volt-amp (VA) rating assigned to the transformer. For example, take a step-down transformer with a primary voltage rating of 120 volts, a secondary voltage rating of 48 volts, and a VA rating of 1 kVA (1000 VA).

The maximum winding currents can be determined as such:

$$\frac{1000\,\text{VA}}{120\,\text{V}} = 8.333\,\text{A}\,(\text{maximum primary winding current})$$

$$\frac{1000\,\text{VA}}{48\,\text{V}} = 20.833\,\text{A}\,(\text{maximum secondary winding current})$$

Sometimes windings will bear current ratings in amps, but this is typically seen on small transformers. Large transformers are almost always rated in terms of winding voltage and VA or kVA.

ENERGY LOSSES

When transformers transfer power, they do so with a minimum of loss. As it was stated earlier, modern power transformer designs typically exceed 95% efficiency. It is good to know where some of this lost power goes, however, and what causes it to be lost. There is, of course, power lost due to resistance of the wire windings. Unless superconducting wires are used, there will always be power dissipated in the form of heat through the resistance of current-carrying conductors. Because transformers require such long lengths of wire, this loss can be a significant factor. Increasing the gauge of the winding wire is one way to minimize this loss, but only with substantial increases in cost, size, and weight.

Resistive losses aside, the bulk of transformer power loss is due to magnetic effects in the core. Perhaps the most significant of these"core losses" is eddy-current loss, which is resistive power dissipation due to the passage of induced currents through the iron of the core. Because iron is a conductor of electricity as well as being an excellent"conductor" of magnetic flux, there will be currents induced in the iron just as there are currents induced in the secondary windings from the alternating magnetic field. These induced currents -- as described by the perpendicularity clause of Faraday's Law -- tend to circulate through the cross-section of the core perpendicularly to the primary winding turns. Their circular motion gives them their unusual name: like eddies in a stream of water that circulate rather than move in straight lines.

Iron is a fair conductor of electricity, but not as good as the copper or aluminum from which wire windings are typically made. Consequently, these"eddy currents" must overcome significant electrical resistance as they circulate through the core. In overcoming the resistance offered by the iron, they dissipate power in the form of heat. Hence, we have a source of inefficiency in the transformer that is difficult to eliminate. This phenomenon is so pronounced that it is often exploited as a means of heating ferrous (iron-containing) materials. The photograph of (Figure below) shows an"induction heating" unit raising the temperature of a large pipe section. Loops of wire covered by high-temperature insulation encircle the pipe's circumference, inducing eddy currents within the pipe wall by electromagnetic induction. In order to maximize the eddy current effect, high-frequency alternating current is used rather than power line frequency (60 Hz). The box units at the right of the picture produce the high-frequency AC and control the amount of current in the wires to stabilize the pipe temperature at a pre-determined"set-point."

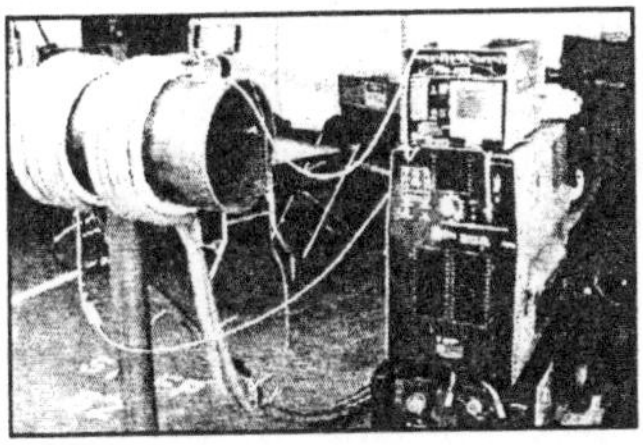

Fig.*Induction Heating*: Primary insulated Winding Induces Current into Lossy iron pipe (secondary).

The main strategy in mitigating these wasteful eddy currents in transformer cores is to form the iron core in sheets, each sheet covered with an insulating varnish so that the core is divided up into thin slices. The result is very little width in the core for eddy currents to circulate in:

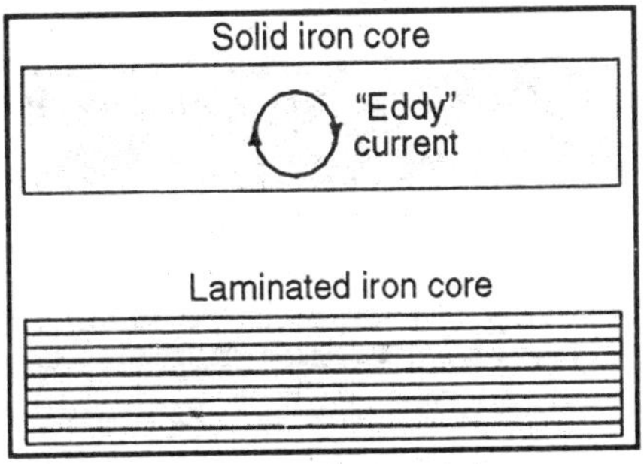

Fig. Dividing the Iron Core into Thin Insulated Laminations Minimizes eddy Current loss.

Laminated cores like the one shown here are standard in almost all low-frequency transformers. Recall from the photograph of the transformer cut in half that the iron core was composed of many thin sheets rather than one solid piece. Eddy current losses increase with frequency, so transformers designed to run on higher-frequency power (such as 400 Hz, used in many military and aircraft applications) must use thinner laminations to keep the losses down to a respectable minimum. This has the undesirable effect of increasing the manufacturing cost of the transformer.

Another, similar technique for minimizing eddy current losses which works better for high-frequency applications is to make the core out of iron powder instead of thin iron sheets. Like the lamination sheets, these granules of iron are individually coated in an electrically insulating material, which makes the core nonconductive except for within the width of each granule. Powdered iron cores are often found in transformers handling radio-frequency currents.

Another"core loss" is that of magnetic hysteresis. All ferromagnetic materials tend to retain some degree of magnetization after exposure to an external magnetic field. This tendency to stay magnetized is called"hysteresis," and it takes a certain investment in energy to overcome this opposition to change every time the magnetic field produced by the primary winding changes polarity (twice per AC cycle). This type of loss can be mitigated through good core material selection (choosing a core alloy with low hysteresis, as evidenced by a"thin" B/H hysteresis curve), and designing the core for minimum flux density (large cross-sectional area).

Transformer energy losses tend to worsen with increasing frequency. The skin effect within winding conductors reduces the available cross-sectional area for electron flow, thereby increasing effective resistance as the frequency goes up and creating more power lost through resistive dissipation. Magnetic core losses are also exaggerated with higher frequencies, eddy currents and hysteresis effects becoming more severe. For this reason, transformers of significant size are designed to operate efficiently in a limited range of frequencies. In most power distribution systems where the line frequency is very stable, one would think excessive frequency would never pose a problem. Unfortunately it does, in the form of harmonics created by nonlinear loads.

As we've seen in earlier chapters, nonsinusoidal waveforms are equivalent to additive series of multiple sinusoidal waveforms at different amplitudes and frequencies. In power systems, these other frequencies are whole-number multiples of the fundamental (line) frequency, meaning that they will always be higher, not lower, than the design frequency of the transformer. In significant measure, they can cause severe transformer overheating. Power transformers can be engineered to handle certain levels of power system harmonics, and this capability is sometimes denoted with a"K factor" rating.

STRAY CAPACITANCE AND INDUCTANCE

Aside from power ratings and power losses, transformers often harbor other undesirable limitations which circuit designers must be made aware of. Like their simpler counterparts -- inductors -- transformers exhibit capacitance due to the insulation dielectric between conductors: from winding to winding, turn to turn (in a single winding), and winding to core. Usually this capacitance is of no concern in a power application, but small signal applications (especially those of high frequency) may not tolerate this quirk well.

Also, the effect of having capacitance along with the windings' designed inductance gives transformers the ability to resonate at a particular frequency, definitely a design concern in signal applications where the applied frequency may reach this point (usually the resonant frequency of a power transformer is well beyond the frequency of the AC power it was designed to operate on).

Flux containment (making sure a transformer's magnetic flux doesn't escape so as to interfere with another device, and making sure other devices' magnetic flux is shielded from the transformer core) is another concern shared both by inductors and transformers.

Closely related to the issue of flux containment is leakage inductance. We've already seen the detrimental effects of leakage inductance on voltage regulation with SPICE simulations early in this chapter. Because leakage inductance is equivalent to an inductance connected in series with the transformer's winding, it manifests itself as a series impedance with the load. Thus, the more current drawn by the load, the less voltage available at the secondary winding terminals. Usually, good voltage regulation is desired in transformer design, but there are exceptional applications.

As was stated before, discharge lighting circuits require a step-up transformer with"loose" (poor) voltage regulation to ensure reduced voltage after the establishment of an arc through the lamp. One way to meet this design criterion is to engineer the transformer with flux leakage paths for magnetic flux to bypass the secondary winding(s). The resulting leakage flux will produce leakage inductance, which will in turn produce the poor regulation needed for discharge lighting.

CORE SATURATION

Transformers are also constrained in their performance by the magnetic flux limitations of the core. For ferromagnetic core transformers, we must be mindful of the saturation limits of the core. Remember that ferromagnetic materials cannot support infinite magnetic flux densities: they tend to"saturate" at a certain level (dictated by the material and core dimensions), meaning that further increases in magnetic field force (mmf) do not result in proportional increases in magnetic field flux (Φ). When a transformer's primary winding is overloaded from excessive applied voltage, the core flux may reach saturation levels during peak moments of the AC sinewave cycle. If this happens, the voltage induced in the secondary winding will no longer match the wave-shape as the voltage powering the primary coil. In other words, the overloaded transformer will distort the waveshape from primary to secondary windings, creating harmonics in the secondary winding's output. As we discussed before, harmonic content in AC power systems typically causes problems.

Special transformers known as peaking transformers exploit this principle to produce brief voltage pulses near the peaks of the source voltage waveform. The core is designed to saturate quickly and sharply, at voltage levels well below peak. This results in a severely cropped sine-wave flux waveform, and secondary voltage pulses only when the flux is changing (below saturation levels):

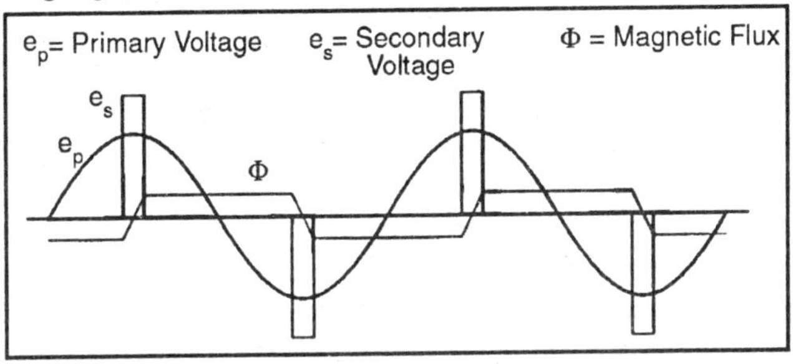

Fig. Voltage and Flux Waveforms for a Peaking Transformer.

Another cause of abnormal transformer core saturation is operation at frequencies lower than normal. For example, if a power transformer designed to operate at 60 Hz is forced to operate at 50 Hz instead, the flux must reach greater peak levels than before in order to produce the same opposing voltage needed to balance against the source voltage. This is true even if the source voltage is the same as before.

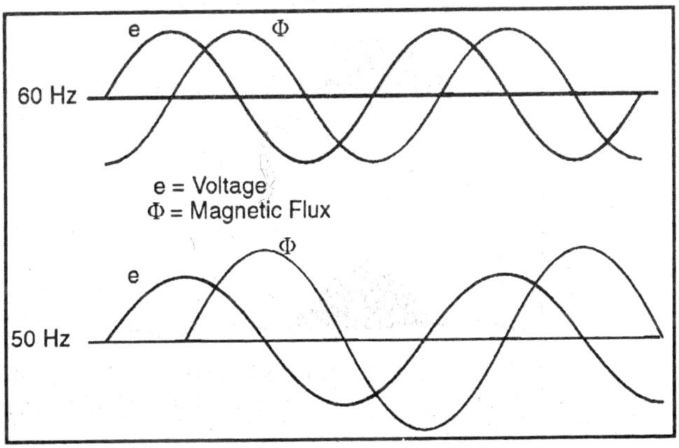

Fig. Magnetic Flux is Higher in a Transformer core Driven by 50 Hz as Compared to 60 Hz for the same Voltage.

Since instantaneous winding voltage is proportional to the instantaneous magnetic flux's rate of change in a transformer, a voltage waveform reaching the same peak value, but taking a longer amount of time to complete each half-cycle, demands that the flux maintain the same rate of change as before, but for longer periods of time. Thus, if the flux has to climb at the same rate as before, but for longer periods of time, it will climb to a greater peak value. Mathematically, this is another example of calculus in action. Because the voltage is proportional to the flux's rate-of-change, we say that the voltage waveform is the derivative of the flux waveform,"derivative" being that calculus operation defining one mathematical function (waveform) in terms of the rate-of-change of another. If we take the opposite perspective, though, and relate the original waveform to its derivative, we may call the original waveform the integral of the derivative waveform. In this case, the voltage waveform is the derivative of the flux waveform, and the flux waveform is the integral of the voltage waveform.

The integral of any mathematical function is proportional to the area accumulated underneath the curve of that function. Since each half-cycle of the 50 Hz waveform accumulates more area between it and the zero line of the graph than the 60 Hz waveform will -- and we know that the magnetic flux is the integral of the voltage -- the flux will attain higher values.

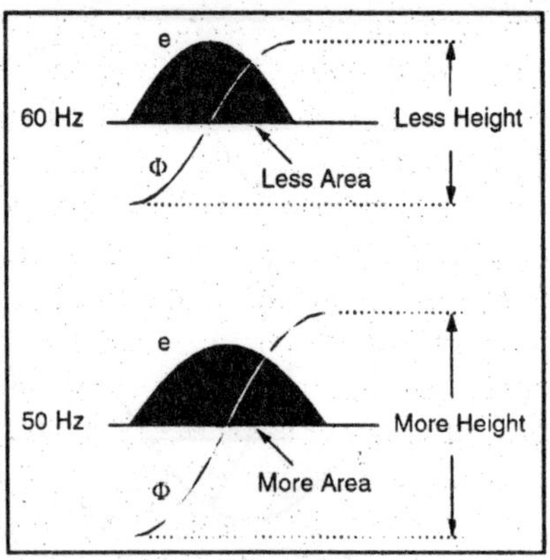

Fig. Flux Changing at the same rate Rises to a Higher level at 50 Hz than at 60 Hz.

Yet another cause of transformer saturation is the presence of DC current in the primary winding. Any amount of DC voltage dropped across the primary winding of a transformer will cause additional magnetic flux in the core.

This additional flux"bias" or"offset" will push the alternating flux waveform closer to saturation in one half-cycle than the other.

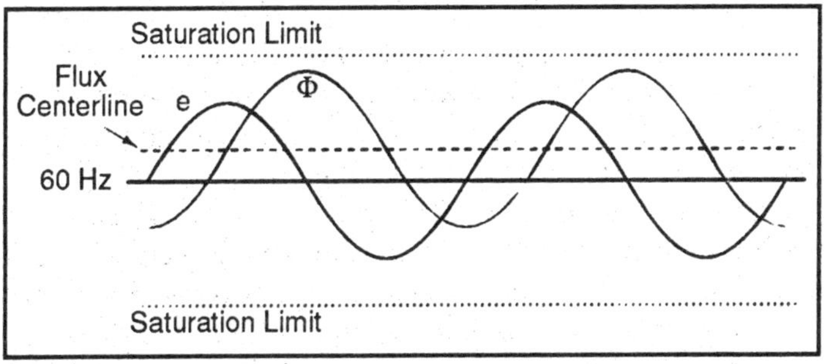

Fig.DC in Primary, Shifts the Waveform peaks Toward the Upper Saturation limit.

For most transformers, core saturation is a very undesirable effect, and it is avoided through good design: engineering the windings and core so that magnetic flux densities remain well below the saturation levels.

This ensures that the relationship between mmf and ? is more linear throughout the flux cycle, which is good because it makes for less distortion in the magnetization current waveform.

Also, engineering the core for low flux densities provides a safe margin between the normal flux peaks and the core saturation limits to accommodate occasional, abnormal conditions such as frequency variation and DC offset.

INRUSH CURRENT

When a transformer is initially connected to a source of AC voltage, there may be a substantial surge of current through the primary winding called inrush current.

This is analogous to the inrush current exhibited by an electric motor that is started up by sudden connection to a power source, although transformer inrush is caused by a different phenomenon.

We know that the rate of change of instantaneous flux in a transformer core is proportional to the instantaneous voltage drop across the primary winding.

Or, as stated before, the voltage waveform is the derivative of the flux waveform, and the flux waveform is the integral of the voltage waveform.

In a continuously-operating transformer, these two waveforms are phase-shifted by 90°. Since flux (Φ) is proportional to the magnetomotive force (mmf) in the core, and the mmf is proportional to winding current, the current waveform will be in-phase with the flux waveform, and both will be lagging the voltage waveform by 90°:

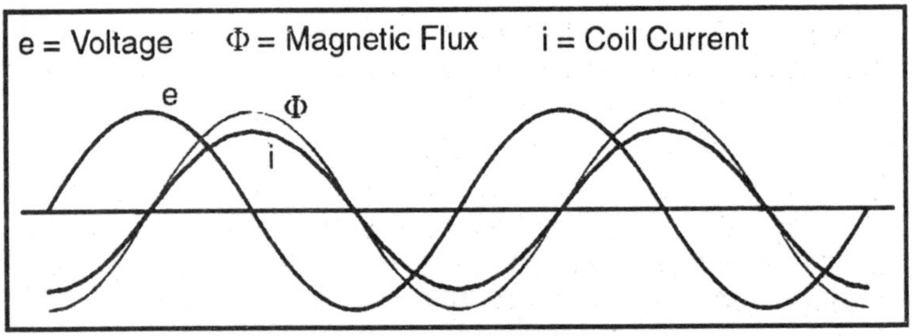

Fig. *Continuous Steady-state Operation*: Magnetic Flux, like Current, lags Applied Voltage by 90°.

Let us suppose that the primary winding of a transformer is suddenly connected to an AC voltage source at the exact moment in time when the instantaneous voltage is at its positive peak value.

In order for the transformer to create an opposing voltage drop to balance against this applied source voltage, a magnetic flux of rapidly increasing value must be generated. The result is that winding current increases rapidly, but actually no more rapidly than under normal conditions:

Both core flux and coil current start from zero and build up to the same peak values experienced during continuous operation. Thus, there is no"surge" or"inrush" or current in this scenario.

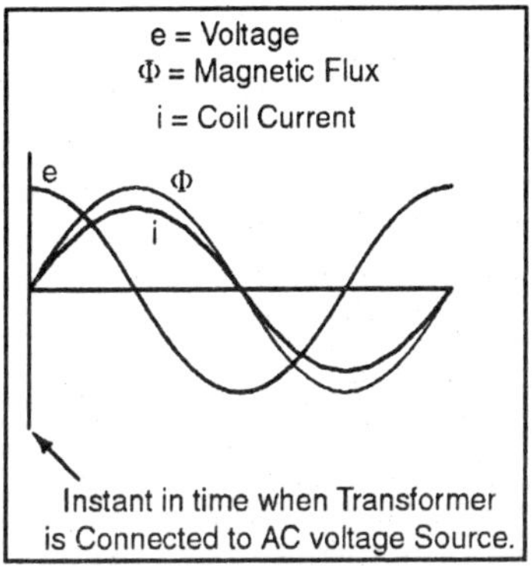

Fig. *Connecting Transformer to Line at AC Volt Peak*: Flux Increases Rapidly from Zero, same as Steady-state Operation.

Alternatively, let us consider what happens if the transformer's connection to the AC voltage source occurs at the exact moment in time when the instantaneous voltage is at zero. During continuous operation (when the transformer has been powered for quite some time), this is the point in time where both flux and winding current are at their negative peaks, experiencing zero rate-of-change ($d\Phi/dt = 0$ and $di/dt = 0$). As the voltage builds to its positive peak, the flux and current waveforms build to their maximum positive rates-of-change, and on upward to their positive peaks as the voltage descends to a level of zero:

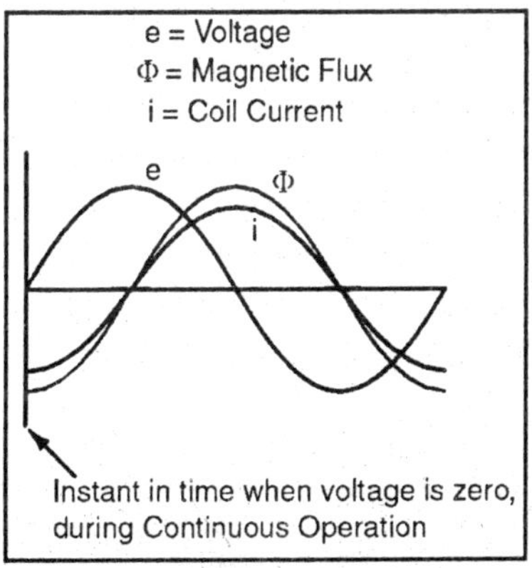

Fig. Starting at e=0 V is not the same as Running Continuously in. These Expected Waveforms are Incorrect- Φ and i should start at zero.

A significant difference exists, however, between continuous-mode operation and the sudden starting condition assumed in this scenario: during continuous operation, the flux and current levels were at their negative peaks when voltage was at its zero point; in a transformer that has been sitting idle, however, both magnetic flux and winding current should start at zero.

When the magnetic flux increases in response to a rising voltage, it will increase from zero upward, not from a previously negative (magnetized) condition as we would normally have in a transformer that's been powered for awhile. Thus, in a transformer that's just"starting," the flux will reach approximately twice its normal peak magnitude as it"integrates" the area under the voltage waveform's first half-cycle:

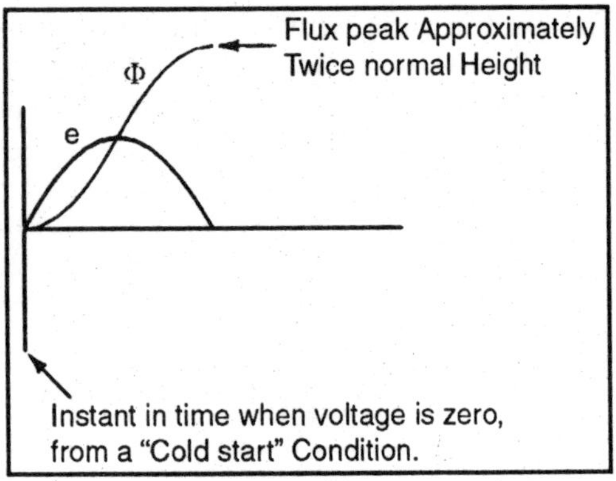

Fig. Starting at e=0 V, Φ starts at Initial Condition Φ=0, Increasing to Twice the Normal value, Assuming it doesn't Saturate the core.

In an ideal transformer, the magnetizing current would rise to approximately twice its normal peak value as well, generating the necessary mmf to create this higher-than-normal flux.

However, most transformers aren't designed with enough of a margin between normal flux peaks and the saturation limits to avoid saturating in a condition like this, and so the core will almost certainly saturate during this first half-cycle of voltage.

During saturation, disproportionate amounts of mmf are needed to generate magnetic flux. This means that winding current, which creates the mmf to cause flux in the core, will disproportionately rise to a value easily exceeding twice its normal peak:

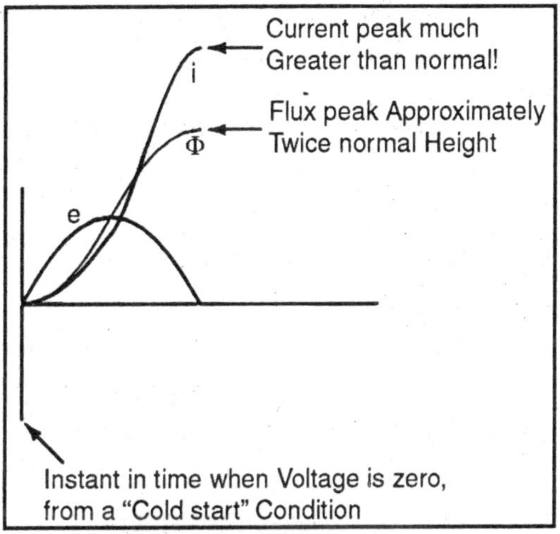

Fig. Starting at e=0 V, Current also Increases to Twice the Normal value for an Unsaturated core, or Considerably Higher in the (Designed for) case of Saturation.

This is the mechanism causing inrush current in a transformer's primary winding when connected to an AC voltage source. As you can see, the magnitude of the inrush current strongly depends on the exact time that electrical connection to the source is made.

If the transformer happens to have some residual magnetism in its core at the moment of connection to the source, the inrush could be even more severe. Because of this, transformer overcurrent protection devices are usually of the"slow-acting" variety, so as to tolerate current surges such as this without opening the circuit.

HEAT AND NOISE

In addition to unwanted electrical effects, transformers may also exhibit undesirable physical effects, the most notable being the production of heat and noise. Noise is primarily a nuisance effect, but heat is a potentially serious problem because winding insulation will be damaged if allowed to overheat.

Heating may be minimized by good design, ensuring that the core does not approach saturation levels, that eddy currents are minimized, and that the windings are not overloaded or operated too close to maximum ampacity.

Large power transformers have their core and windings submerged in an oil bath to transfer heat and muffle noise, and also to displace moisture which would otherwise compromise the integrity of the winding insulation. Heat-dissipating"radiator" tubes on the outside of the transformer case provide a convective oil flow path to transfer heat from the transformer's core to ambient air: (Figure below)

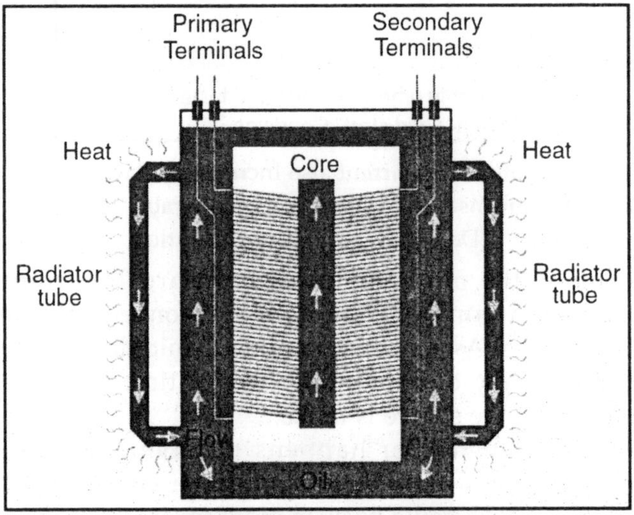

Fig. Large Power Transformers are Submerged in Heat Dissipating Insulating oil.

Oil-less, or"dry," transformers are often rated in terms of maximum operating temperature"rise" (temperature increase beyond ambient) according to a letter-class system. *A, B, F, or H. These letter codes are arranged in order of lowest heat tolerance to highest*:

- *Class A*: No more than 55° Celsius winding temperature rise, at 40° Celsius (maximum) ambient air temperature.
- *Class B*: No more than 80° Celsius winding temperature rise, at 40° Celsius (maximum)ambient air temperature.
- *Class F*: No more than 115° Celsius winding temperature rise, at 40° Celsius (maximum)ambient air temperature.
- *Class H*: No more than 150° Celsius winding temperature rise, at 40° Celsius (maximum)ambient air temperature.

Audible noise is an effect primarily originating from the phenomenon of magnetostriction: the slight change of length exhibited by a ferromagnetic object when magnetized. The familiar"hum" heard around large power transformers is the sound of the iron core expanding and contracting at 120 Hz (twice the system frequency, which is 60 Hz in the United States) -- one cycle of core contraction and expansion for every peak of the magnetic flux waveform -- plus noise created by mechanical forces between primary and secondary windings. Again, maintaining low magnetic flux levels in the core is the key to minimizing this effect, which explains why ferroresonant transformers -- which must operate in saturation for a large portion of the current waveform -- operate both hot and noisy.

Another noise-producing phenomenon in power transformers is the physical reaction force between primary and secondary windings when heavily loaded. If the secondary winding is open-circuited, there will be no current through it, and consequently no magneto-motive force (mmf) produced by it. However, when the secondary is"loaded" (current supplied to a load), the winding generates an mmf, which becomes counteracted by a"reflected" mmf in the primary winding to prevent core flux levels from changing.

These opposing mmf's generated between primary and secondary windings as a result of secondary (load) current produce a repulsive, physical force between the windings which will tend to make them vibrate. Transformer designers have to consider these physical forces in the construction of the winding coils, to ensure there is adequate mechanical support to handle the stresses. Under heavy load (high current) conditions, though, these stresses may be great enough to cause audible noise to emanate from the transformer.

Chapter 6

Role of Switches and Multivibrators

ELECTRICAL SWITCH

An electrical switch is any device used to interrupt the flow of electrons in a circuit. Switches are essentially binary devices: they are either completely on ("closed") or completely off ("open"). There are many different types of switches, and we will explore some of these types in this chapter.

Though it may seem strange to cover this elementary electrical topic at such a late stage in this book series, I do so because the chapters that follow explore an older realm of digital technology based on mechanical switch contacts rather than solid-state gate circuits, and a thorough understanding of switch types is necessary for the undertaking. Learning the function of switch-based circuits at the same time that you learn about solid-state logic gates makes both topics easier to grasp, and sets the stage for an enhanced learning experience in Boolean algebra, the mathematics behind digital logic circuits.

The simplest type of switch is one where two electrical conductors are brought in contact with each other by the motion of an actuating mechanism.

Other switches are more complex, containing electronic circuits able to turn on or off depending on some physical stimulus (such as light or magnetic field) sensed. In any case, the final output of any switch will be (at least) a pair of wire-connection terminals that will either be connected together by the switch's internal contact mechanism ("closed"), or not connected together ("open").

Any switch designed to be operated by a person is generally called a *hand switch*, and they are manufactured in several varieties:

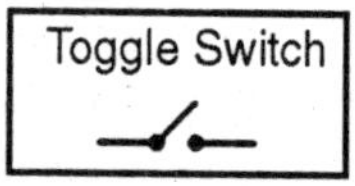

Toggle switches are actuated by a lever angled in one of two or more positions. The common light switch used in household wiring is an example of a toggle switch.

Most toggle switches will come to rest in any of their lever positions, while others have an internal spring mechanism returning the lever to a certain *normal* position, allowing for what is called "momentary" operation.

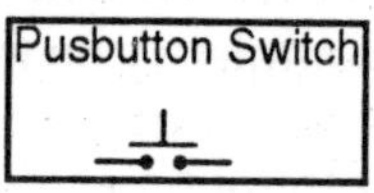

Pushbutton switches are two-position devices actuated with a button that is pressed and released. Most pushbutton switches have an internal spring mechanism returning the button to its "out," or "unpressed," position, for momentary operation.

Some pushbutton switches will latch alternately on or off with every push of the button. Other pushbutton switches will stay in their "in," or "pressed," position until the button is pulled back out. This last type of pushbutton switches usually have a mushroom-shaped button for easy push-pull action.

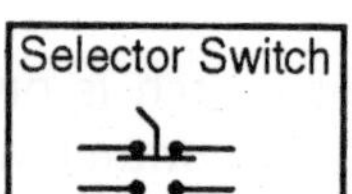

Selector switches are actuated with a rotary knob or lever of some sort to select one of two or more positions. Like the toggle switch, selector switches can either rest in any of their positions or contain spring-return mechanisms for momentary operation.

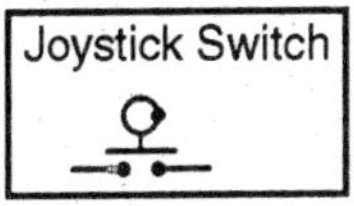

A joystick switch is actuated by a lever free to move in more than one axis of motion. One or more of several switch contact mechanisms are actuated depending on which way the lever is pushed, and sometimes by how *far* it is pushed. The circle-and-dot notation on the switch symbol represents the direction of joystick lever motion required to actuate the contact. Joystick hand switches are commonly used for crane and robot control.

Some switches are specifically designed to be operated by the motion of a machine rather than by the hand of a human operator. These motion-operated switches are commonly called *limit switches,* because they are often used to limit the motion of a machine by turning off the actuating power to a component if it moves too far. As with hand switches, limit switches come in several varieties:

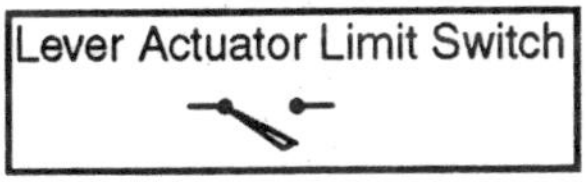

These limit switches closely resemble rugged toggle or selector hand switches fitted with a lever pushed by the machine part. Often, the levers are tipped with a small roller bearing, preventing the lever from being worn off by repeated contact with the machine part.

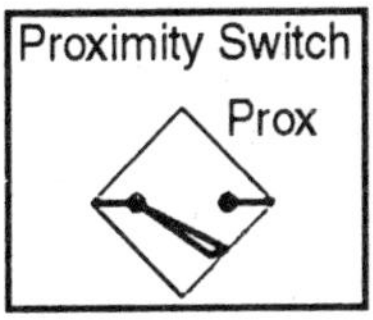

Proximity switches sense the approach of a metallic machine part either by a magnetic or high-frequency electromagnetic field. Simple proximity switches use a permanent magnet to actuate a sealed switch mechanism whenever the machine part gets close (typically 1 inch or less). More complex proximity switches work like a metal detector, energizing a coil of wire with a high-frequency current, and electronically monitoring the magnitude of that current.

If a metallic part (not necessarily magnetic) gets close enough to the coil, the current will increase, and trip the monitoring circuit. The symbol shown here for the proximity switch is of the electronic variety, as indicated by the diamond-shaped box surrounding the switch. A non-electronic proximity switch would use the same symbol as the lever-actuated limit switch.

Another form of proximity switch is the optical switch, comprised of a light source and photocell. Machine position is detected by either the interruption or reflection of a light beam. Optical switches are also useful in safety applications, where beams of light can be used to detect personnel entry into a dangerous area. In many industrial processes, it is necessary to monitor various physical quantities with switches. Such switches can be used to sound alarms, indicating that a process variable has exceeded normal parametres, or they can be used to shut down processes or equipment if those variables have reached dangerous or destructive levels.

There are many different types of process switches:

These switches sense the rotary speed of a shaft either by a centrifugal weight mechanism mounted on the shaft, or by some kind of non-contact detection of shaft motion such as optical or magnetic.

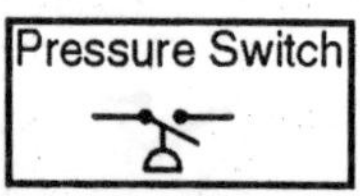

Gas or liquid pressure can be used to actuate a switch mechanism if that pressure is applied to a piston, diaphragm, or bellows, which converts pressure to mechanical force.

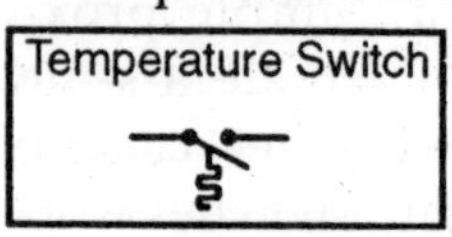

An inexpensive temperature-sensing mechanism is the "bimetallic strip:" a thin strip of two metals, joined back-to-back, each metal having a different rate of thermal expansion. When the strip heats or cools, differing rates of thermal expansion between the two metals causes it to bend. The bending of the strip can then be used to actuate a switch contact mechanism. Other temperature switches use a brass bulb filled with either a liquid or gas, with a tiny tube connecting the bulb to a pressure-sensing switch. As the bulb is heated, the gas or liquid expands, generating a pressure increase which then actuates the switch mechanism.

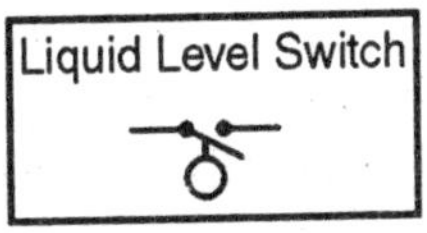

A floating object can be used to actuate a switch mechanism when the liquid level in an tank rises past a certain point. If the liquid is electrically conductive, the liquid itself can be used as a conductor to bridge between two metal probes inserted into the tank at the required depth. The conductivity technique is usually implemented with a special design of relay triggered by a small amount of current through the conductive liquid. In most cases it is impractical and dangerous to switch the full load current of the circuit through a liquid.

Level switches can also be designed to detect the level of solid materials such as wood chips, grain, coal, or animal feed in a storage silo, bin, or hopper. A common design for this application is a small paddle wheel, inserted into the bin at the desired height, which is slowly turned by a small electric motor.

When the solid material fills the bin to that height, the material prevents the paddle wheel from turning. The torque response of the small motor than trips the switch mechanism. Another design uses a "tuning fork" shaped metal prong, inserted into the bin from the outside at the desired height. The fork is vibrated at its resonant frequency by an electronic circuit and magnet/electromagnet coil assembly. When the bin fills to that height, the solid material dampens the vibration of the fork, the change in vibration amplitude and/or frequency detected by the electronic circuit.

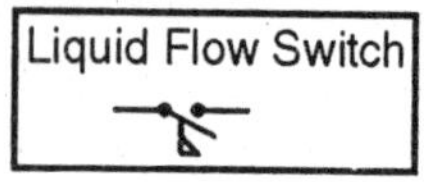

Inserted into a pipe, a flow switch will detect any gas or liquid flow rate in excess of a certain threshold, usually with a small paddle or vane which is pushed by the flow. Other flow switches are constructed as differential pressure switches, measuring the pressure drop across a restriction built into the pipe.

Another type of level switch, suitable for liquid or solid material detection, is the nuclear switch. Composed of a radioactive source material and a radiation detector, the two are mounted across the diametre of a storage vessel for either solid or liquid material. Any height of material beyond the level of the source/detector arrangement will attenuate the strength of radiation reaching the detector. This decrease in radiation at the detector can be used to trigger a relay mechanism to provide a switch contact for measurement, alarm point, or even control of the vessel level.

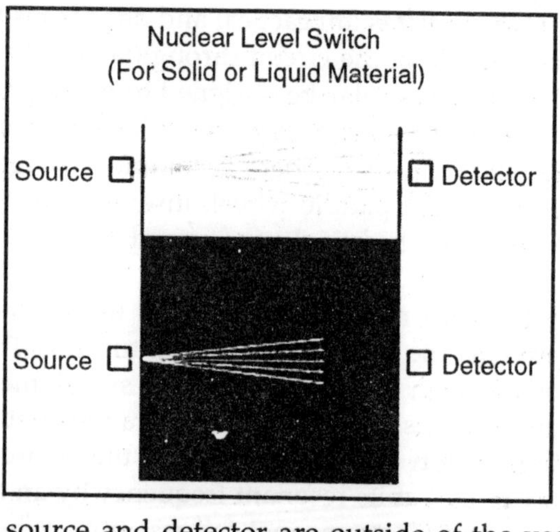

Both source and detector are outside of the vessel, with no intrusion at all except the radiation flux itself. The radioactive sources used are fairly weak and pose no immediate health threat to operations or maintenance personnel.

As usual, there is usually more than one way to implement a switch to monitor a physical process or serve as an operator control. There is usually no single "perfect" switch for any application, although some obviously exhibit certain advantages over others. Switches must be intelligently matched to the task for efficient and reliable operation.

DESIGN OF SWITCH

A switch can be constructed with any mechanism bringing two conductors into contact with each other in a controlled manner. This can be as simple as allowing two copper wires to touch each other by the motion of a lever, or by directly pushing two metal strips into contact. However, a good switch design must be rugged and reliable, and avoid presenting the operator with the possibility of electric shock. Therefore, industrial switch designs are rarely this crude.

The conductive parts in a switch used to make and break the electrical connection are called *contacts*. Contacts are typically made of silver or silver-cadmium alloy, whose conductive properties are not significantly compromised by surface corrosion or oxidation. Gold contacts exhibit the best corrosion resistance, but are limited in current-carrying capacity and may "cold weld" if brought together with high mechanical force. Whatever the choice of metal, the switch contacts are guided by a mechanism ensuring square and even contact, for maximum reliability and minimum resistance.

Contacts such as these can be constructed to handle extremely large amounts of electric current, up to thousands of amps in some cases.

The limiting factors for switch contact ampacity are as follows:

- Heat generated by current through metal contacts (while closed).
- Sparking caused when contacts are opened or closed.
- The voltage across open switch contacts (potential of current jumping across the gap).

One major disadvantage of standard switch contacts is

the exposure of the contacts to the surrounding atmosphere. In a nice, clean, control-room environment, this is generally not a problem. However, most industrial environments are not this benign. The presence of corrosive chemicals in the air can cause contacts to deteriorate and fail prematurely. Even more troublesome is the possibility of regular contact sparking causing flammable or explosive chemicals to ignite.

When such environmental concerns exist, other types of contacts can be considered for small switches. These other types of contacts are sealed from contact with the outside air, and therefore do not suffer the same exposure problems that standard contacts do.

A common type of sealed-contact switch is the mercury switch. Mercury is a metallic element, liquid at room temperature. Being a metal, it possesses excellent conductive properties. Being a liquid, it can be brought into contact with metal probes (to close a circuit) inside of a sealed chamber simply by tilting the chamber so that the probes are on the bottom.

Many industrial switches use small glass tubes containing mercury which are tilted one way to close the contact, and tilted another way to open. Aside from the problems of tube breakage and spilling mercury (which is a toxic material), and susceptibility to vibration, these devices are an excellent alternative to open-air switch contacts wherever environmental exposure problems are a concern.

FORCE OF ACTUATION

When a switch is actuated and contacts touch one another under the force of actuation, they are supposed to establish continuity in a single, crisp moment. Unfortunately, though, switches do not exactly achieve this goal. Due to the mass of the moving contact and any elasticity inherent in the mechanism and/or contact materials, contacts will "bounce" upon closure for a period of milliseconds before coming to a full rest and providing unbroken contact.

In many applications, switch bounce is of no consequence: it matters little if a switch controlling an incandescent lamp "bounces" for a few cycles every time it is actuated. Since the lamp's warm-up time greatly exceeds the bounce period, no irregularity in lamp operation will result.

However, if the switch is used to send a signal to an electronic amplifier or some other circuit with a fast response time, contact bounce may produce very noticeable and undesired effects:

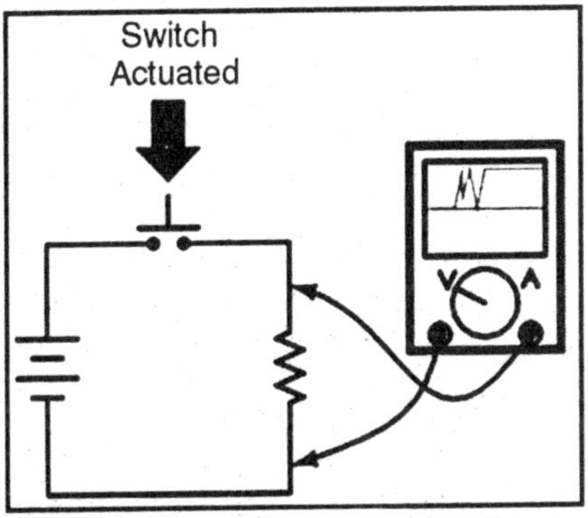

A closer look at the oscilloscope display reveals a rather ugly set of makes and breaks when the switch is actuated a single time:

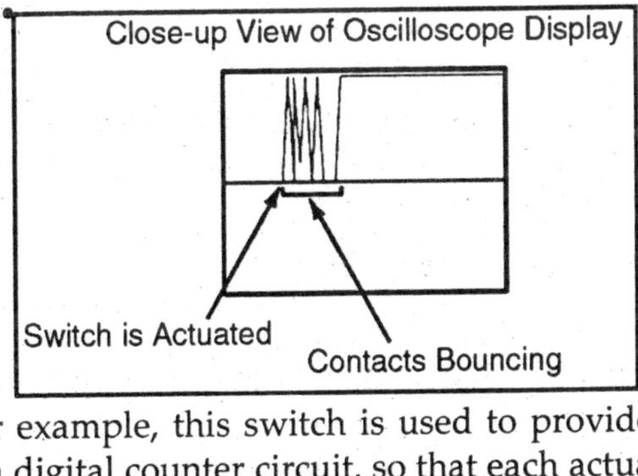

If, for example, this switch is used to provide a "clock" signal to a digital counter circuit, so that each actuation of the pushbutton switch is supposed to increment the counter by a value of 1, what will happen instead is the counter will increment by several counts each time the switch is actuated. Since mechanical switches often interface with digital electronic circuits in modern systems, switch contact bounce is a frequent design consideration. Somehow, the "chattering"

produced by bouncing contacts must be eliminated so that the receiving circuit sees a clean, crisp off/on transition:

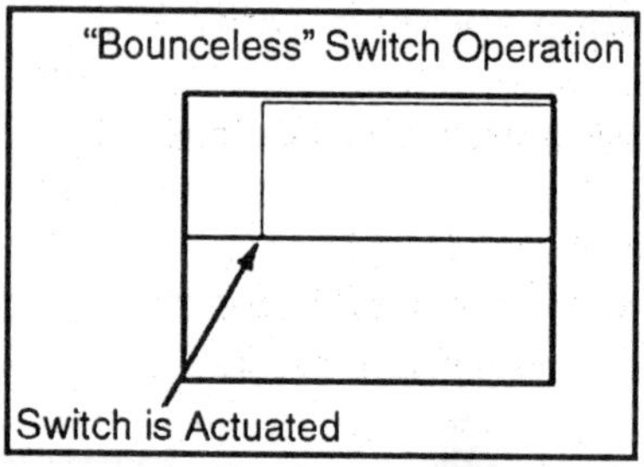

Switch contacts may be *debounced* several different ways. The most direct means is to address the problem at its source: the switch itself. Here are some suggestions for designing switch mechanisms for minimum bounce:

- Reduce the kinetic energy of the moving contact. This will reduce the force of impact as it comes to rest on the stationary contact, thus minimizing bounce.
- Use "buffer springs" on the stationary contact(s) so that they are free to recoil and gently absorb the force of impact from the moving contact.
- Design the switch for "wiping" or "sliding" contact rather than direct impact. "Knife" switch designs use sliding contacts.
- Dampen the switch mechanism's movement using an air or oil "shock absorber" mechanism.
- Use sets of contacts in parallel with each other, each slightly different in mass or contact gap, so that when one is rebounding off the stationary contact, at least one of the others will still be in firm contact.
- "Wet" the contacts with liquid mercury in a sealed environment. After initial contact is made, the surface tension of the mercury will maintain circuit continuity even though the moving contact may bounce off the stationary contact several times.

Each one of these suggestions sacrifices some aspect of switch performance for limited bounce, and so it is impractical to design *all* switches with limited contact bounce in mind. Alterations made to reduce the kinetic energy of the contact may result in a small open-contact gap or a slow-moving contact, which limits the amount of voltage the switch may handle and the amount of current it may interrupt. Sliding contacts, while non-bouncing, still produce "noise" (irregular current caused by irregular contact resistance when moving), and suffer from more mechanical wear than normal contacts.

Multiple, parallel contacts give less bounce, but only at greater switch complexity and cost. Using mercury to "wet" the contacts is a very effective means of bounce mitigation, but it is unfortunately limited to switch contacts of low ampacity. Also, mercury-wetted contacts are usually limited in mounting position, as gravity may cause the contacts to "bridge" accidently if oriented the wrong way.

If re-designing the switch mechanism is not an option, mechanical switch contacts may be debounced externally, using other circuit components to condition the signal. A low-pass filter circuit attached to the output of the switch, for example, will reduce the voltage/current fluctuations generated by contact bounce:

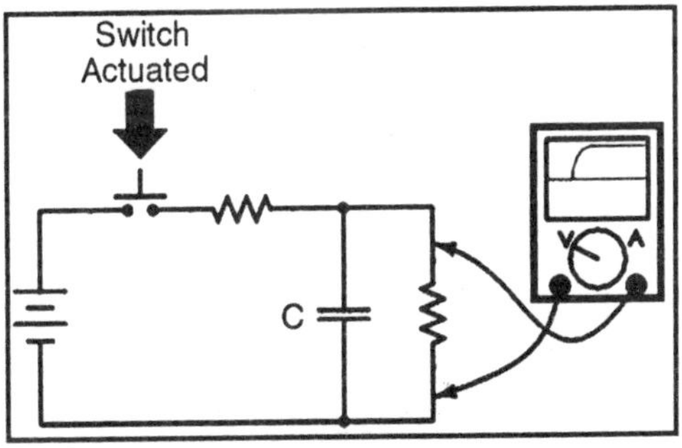

Switch contacts may be debounced electronically, using hysteretic transistor circuits (circuits that "latch" in either a high or a low state) with built-in time delays (called "one-shot" circuits), or two inputs controlled by a double-throw switch.

MULTIVIBRATORS

ROLE OF LOGIC GATE

With simple gate and combinational logic circuits, there is a definite output state for any given input state. Take the truth table of an OR gate, for instance:

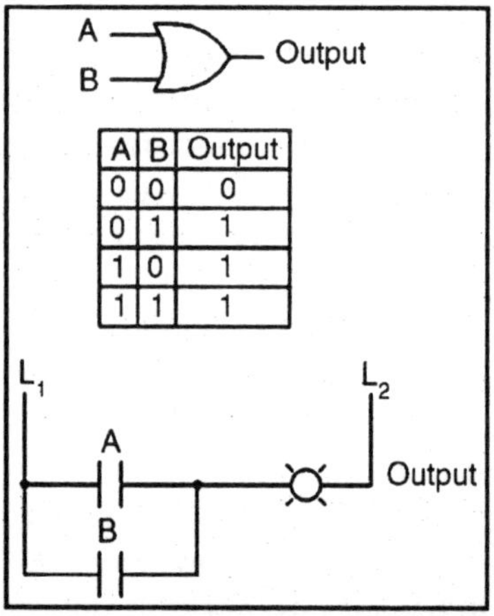

A	B	Output
0	0	0
0	1	1
1	0	1
1	1	1

For each of the four possible combinations of input states (0-0, 0-1, 1-0, and 1-1), there is one, definite, unambiguous output state. Whether we're dealing with a multitude of cascaded gates or a single gate, that output state is determined by the truth table(s) for the gate(s) in the circuit, and nothing else. However, if we alter this gate circuit so as to give signal feedback from the output to one of the inputs, strange things begin to happen:

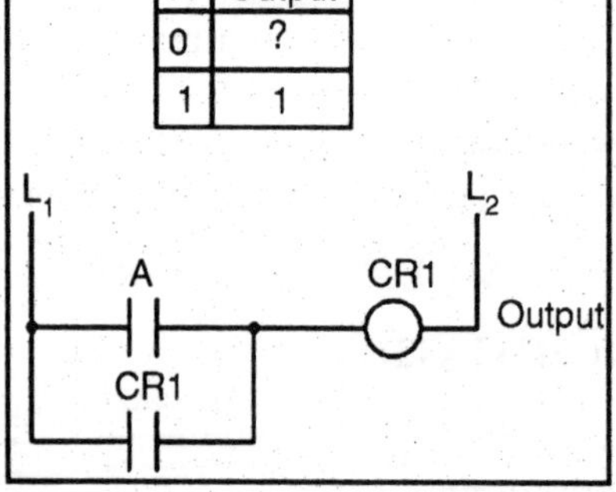

A	Output
0	?
1	1

We know that if A is 1, the output *must* be 1, as well.

Such is the nature of an OR gate: any "high" (1) input forces the output "high" (1). If A is "low" (0), however, we cannot guarantee the logic level or state of the output in our truth table. Since the output feeds back to one of the OR gate's inputs, and we know that any 1 input to an OR gates makes the output 1, this circuit will "latch" in the 1 output state after any time that A is 1. When A is 0, the output could be either 0 or 1,*depending on the circuit's prior state!* The proper way to complete the above truth table would be to insert the word *latch* in place of the question mark, showing that the output maintains its last state when A is 0.

Any digital circuit employing feedback is called a *multivibrator*. The example we just explored with the OR gate was a very simple example of what is called a *bistable* multivibrator. It is called "bistable" because it can hold stable in one of *two*possible output states, either 0 or 1. There are also *monostable* multivibrators, which have only *one* stable output state (that other state being momentary), which we'll explore later; and *astable* multivibrators, which have no stable state (oscillating back and forth between an output of 0 and 1).

A very simple astable multivibrator is an inverter with the output fed directly back to the input:

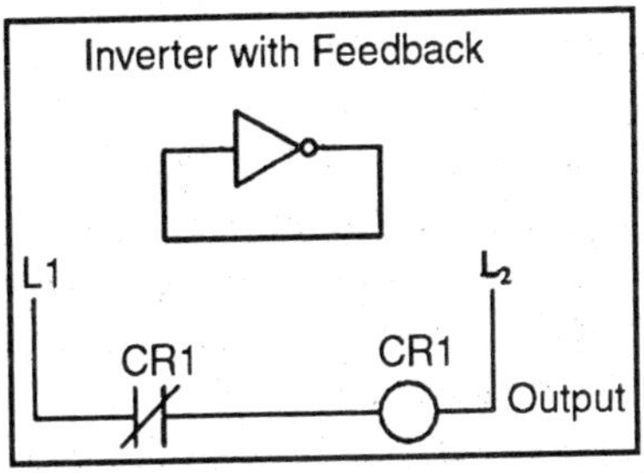

When the input is 0, the output switches to 1. That 1 output gets fed back to the input as a 1. When the input is 1, the output switches to 0. That 0 output gets fed back to the input as a 0, and the cycle repeats itself. The result is a high frequency (several megahertz) oscillator, if implemented with a solid-state (semiconductor) inverter gate:

If implemented with relay logic, the resulting oscillator

will be considerably slower, cycling at a frequency well within the audio range. The *buzzer* or *vibrator* circuit thus formed was used extensively in early radio circuitry, as a way to convert steady, low-voltage DC power into pulsating DC power which could then be stepped up in voltage through a transformer to produce the high voltage necessary for operating the vacuum tube amplifiers. Henry Ford's engineers also employed the buzzer/transformer circuit to create continuous high voltage for operating the spark plugs on Model T automobile engines:

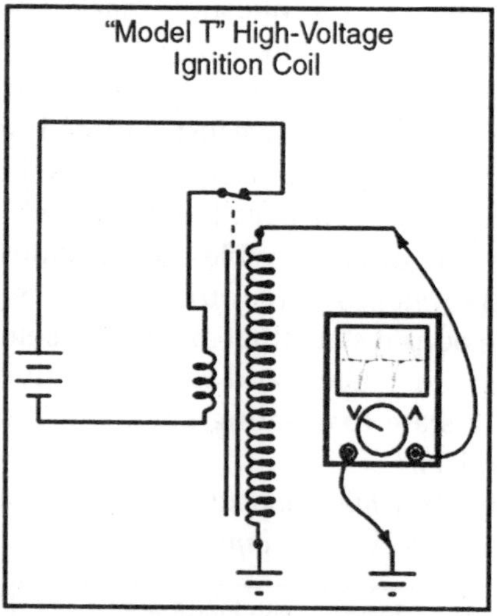

Borrowing terminology from the old mechanical buzzer (vibrator) circuits, solid-state circuit engineers referred to any circuit with two or more vibrators linked together as a *multivibrator*. The astable multivibrator mentioned previously, with only one "vibrator," is more commonly implemented with multiple gates.

The most interesting and widely used multivibrators are of the bistable variety, so we'll explore them in detail now.

STABLE STAGES

The S-R Latch

A bistable multivibrator has *two* stable states, as indicated by the prefix *bi* in its name. Typically, one state is referred to asset and the other as *reset*. The simplest bistable device, therefore, is known as a *set-reset*, or S-R, latch. To create an S-R latch, we can wire two NOR gates in such a way that the output of one feeds back to the input of another, and visa-versa, like this:

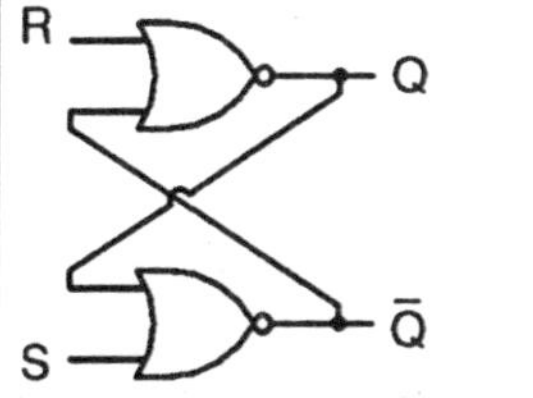

S	R	Q	$\bar{Q}$
0	0	Latch	Latch
0	1	0	1
1	0	1	0
1	1	0	0

The Q and not-Q outputs are supposed to be in opposite states. I say "supposed to" because making both the S and R inputs equal to 1 results in both Q and not-Q being 0. For this reason, having both S and R equal to 1 is called an *invalid* or*illegal* state for the S-R multivibrator. Otherwise, making S=1 and R=0 "sets" the multivibrator so that Q=1 and not-Q=0. Conversely, making R=1 and S=0 "resets" the multivibrator in the opposite state. When S and R are both equal to 0, the multivibrator's outputs "latch" in their prior states.

Note how the same multivibrator function can be implemented in ladder logic, with the same results:

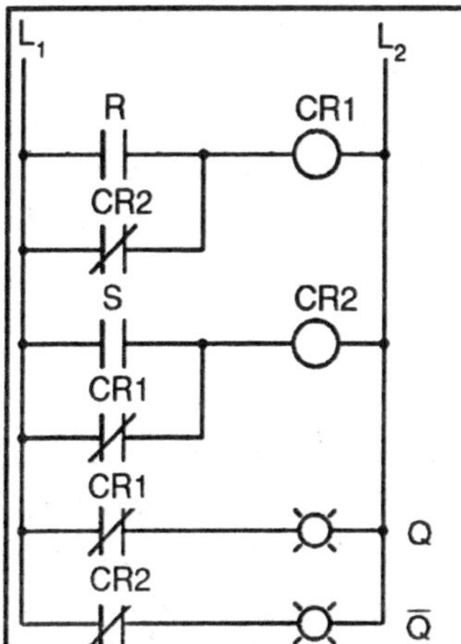

S	R	Q	$\bar{Q}$
0	0	Latch	Latch
0	1	0	1
1	0	1	0
1	1	0	0

By definition, a condition of Q=1 and not-Q=0 is *set*. A

condition of Q=0 and not-Q=1 is *reset*. These terms are universal in describing the output states of any multivibrator circuit.

The astute observer will note that the initial power-up condition of either the gate or ladder variety of S-R latch is such that both gates (coils) start in the de-energized mode. As such, one would expect that the circuit will start up in an invalid condition, with both Q and not-Q outputs being in the same state. Actually, this is true.

However, the invalid condition is unstable with both S and R inputs inactive, and the circuit will quickly stabilize in either the set or reset condition because one gate (or relay) is bound to react a little faster than the other. If both gates (or coils) were *precisely identical,* they would oscillate between high and low like an astable multivibrator upon power-up without ever reaching a point of stability! Fortunately for cases like this, such a precise match of components is a rare possibility.

It must be noted that although an astable (continually oscillating) condition would be extremely rare, there will most likely be a cycle or two of oscillation in the above circuit, and the final state of the circuit (set or reset) after power-up would be unpredictable. The root of the problem is a *race condition* between the two relays CR_1 and CR_2.

A race condition occurs when two mutually-exclusive events are simultaneously initiated through different circuit elements by a single cause. In this case, the circuit elements are relays CR_1 and CR_2, and their de-energized states are mutually exclusive due to the normally-closed interlocking contacts. If one relay coil is de-energized, its normally-closed contact will keep the other coil energized, thus maintaining the circuit in one of two states (set or reset). Interlocking prevents *both*relays from latching.

However, if *both* relay coils start in their de-energized states (such as after the whole circuit has been de-energized and is then powered up) both relays will "race" to become latched on as they receive power (the "single cause") through the normally-closed contact of the other relay.

One of those relays will inevitably reach that condition

before the other, thus opening its normally-closed interlocking contact and de-energizing the other relay coil. Which relay "wins" this race is dependent on the physical characteristics of the relays and not the circuit design, so the designer cannot ensure which state the circuit will fall into after power-up.

Race conditions should be avoided in circuit design primarily for the unpredictability that will be created. One way to avoid such a condition is to insert a time-delay relay into the circuit to disable one of the competing relays for a short time, giving the other one a clear advantage. In other words, by purposely slowing down the de-energization of one relay, we ensure that the other relay will always "win" and the race results will always be predictable. Here is an example of how a time-delay relay might be applied to the above circuit to avoid the race condition:

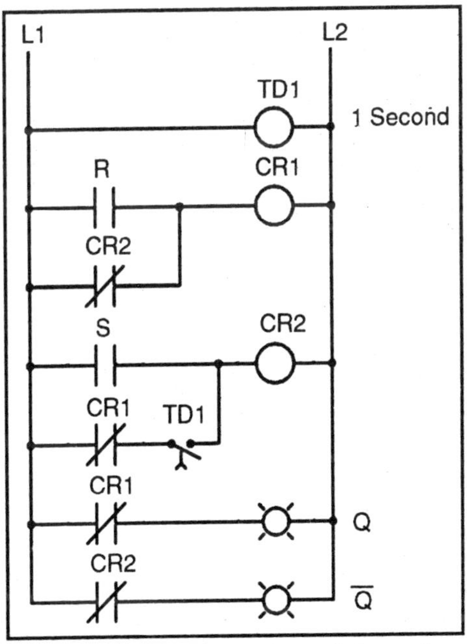

When the circuit powers up, time-delay relay contact TD1 in the fifth rung down will delay closing for 1 second. Having that contact open for 1 second prevents relay CR2 from energizing through contact CR1 in its normally-closed state after power-up. Therefore, relay CR1 will be allowed to energize first (with a 1-second head start), thus opening the normally-closed CR1contact in the fifth rung, preventing CR2 from being energized without the S input going active. The end result is that the circuit powers up cleanly and predictably in the reset state with S=0 and R=0.

It should be mentioned that race conditions are not restricted to relay circuits. Solid-state logic gate circuits may also suffer from the ill effects of race conditions if improperly designed. Complex computer programme, for that matter, may also incur race problems if improperly designed. Race problems are a possibility for any sequential system, and may not be discovered until some time after initial testing of the system.

They can be very difficult problems to detect and eliminate. A practical application of an S-R latch circuit might be for starting and stopping a motor, using normally-open, momentary pushbutton switch contacts for both start (S) and stop (R) switches, then energizing a motor contactor with either a CR1 or CR2 contact (or using a contactor in place of CR1 or CR2).

Normally, a much simpler ladder logic circuit is employed, such as this:

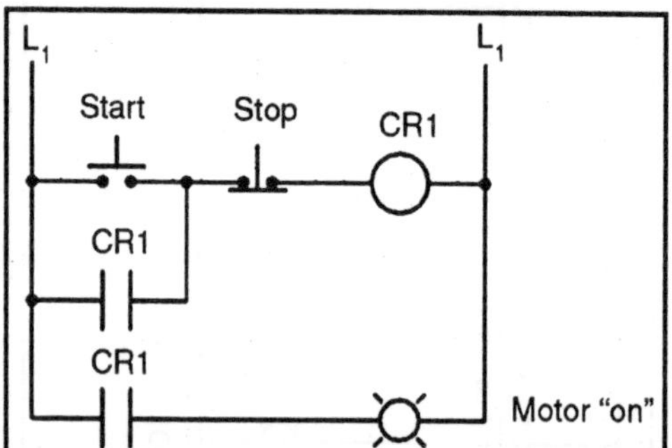

In the above motor start/stop circuit, the CR1 contact in parallel with the start switch contact is referred to as a"seal-in" contact, because it"seals" or latches control relay CR1 in the energized state after the start switch has been released. To break the"seal," or to"unlatch" or"reset" the circuit, the stop pushbutton is pressed, which de-energizes CR1 and restores the seal-in contact to its normally open status. Notice, however, that this circuit performs much the same function as the S-R latch. Also note that this circuit has no inherent instability problem (if even a remote possibility) as does the double-relay S-R latch design.

In semiconductor form, S-R latches come in prepackaged units so that you don't have to build them from individual gates.

They are symbolized as such:

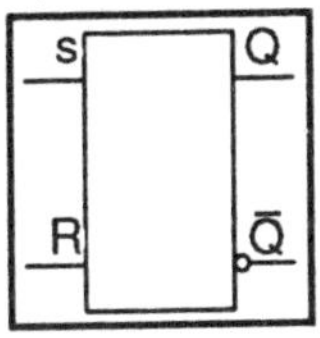

The Gated S-R Latch

It is sometimes useful in logic circuits to have a multivibrator which changes state only when certain conditions are met, regardless of its S and R input states. The conditional input is called the enable, and is symbolized by the letter E. Study the following example:

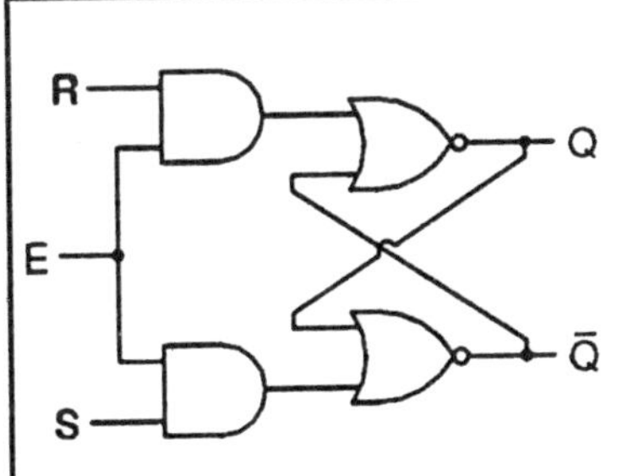

E	S	R	Q	Q̄
0	0	0	Latch	Latch
0	0	1	Latch	Latch
0	1	0	Latch	Latch
0	1	1	Latch	Latch
1	0	0	Latch	Latch
1	0	1	0	1
1	1	0	1	0
1	1	1	0	0

When the E=0, the outputs of the two AND gates are forced to 0, regardless of the states of either S or R. Consequently, the circuit behaves as though S and R were both 0, latching the Q and not-Q outputs in their last states. Only when the enable input is activated (1) will the latch respond to the S and R inputs.

Note the identical function in ladder logic:

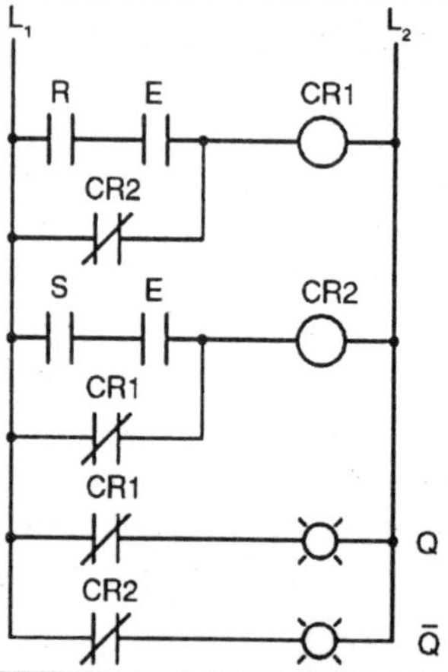

E	S	R	Q	$\bar{Q}$
0	0	0	Latch	Latch
0	0	1	Latch	Latch
0	1	0	Latch	Latch
0	1	1	Latch	Latch
1	0	0	Latch	Latch
1	0	1	0	1
1	1	0	1	0
1	1	1	0	0

A practical application of this might be the same motor control circuit (with two normally-open pushbutton switches for *start*and *stop*), except with the addition of a master lockout input (E) that disables both pushbuttons from having control over the motor when it's low (0). Once again, these multivibrator circuits are available as prepackaged semiconductor devices, and are symbolized as such:

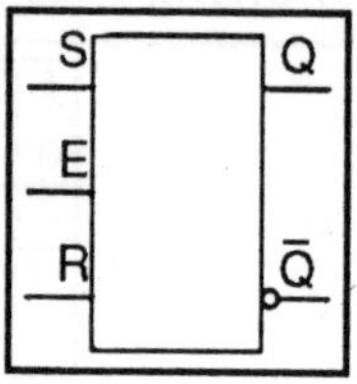

It is also common to see the enable input designated by the letters "EN" instead of just "E."

THE D LATCH

Since the enable input on a gated S-R latch provides a way to latch the Q and not-Q outputs without regard to the status of S or R, we can eliminate one of those inputs to create a multivibrator latch circuit with no "illegal" input states. Such a circuit is called a D latch, and its internal logic looks like this:

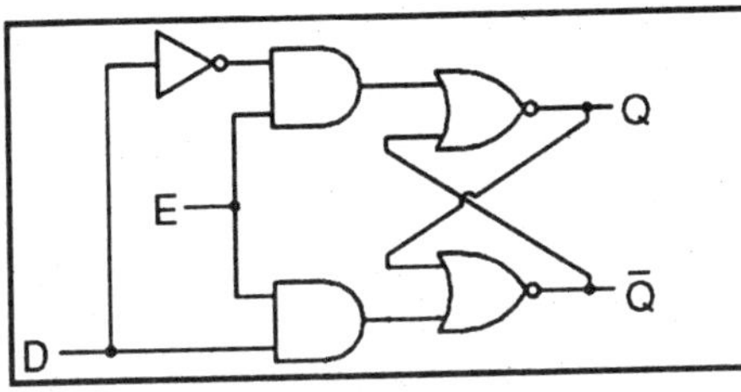

E	D	Q	$\bar{Q}$
0	0	Latch	Latch
0	1	Latch	Latch
1	0	0	1
1	1	1	0

Note that the R input has been replaced with the complement (inversion) of the old S input, and the S input has been renamed to D. As with the gated S-R latch, the D latch will not respond to a signal input if the enable input is 0 — it simply stays latched in its last state. When the enable input is 1, however, the Q output follows the D input. Since the R input of the S-R circuitry has been done away with, this latch has no "invalid" or "illegal" state. Q and not-Q are*always* opposite of one another. If the above diagram is confusing at all, the next diagram should make the concept simpler:

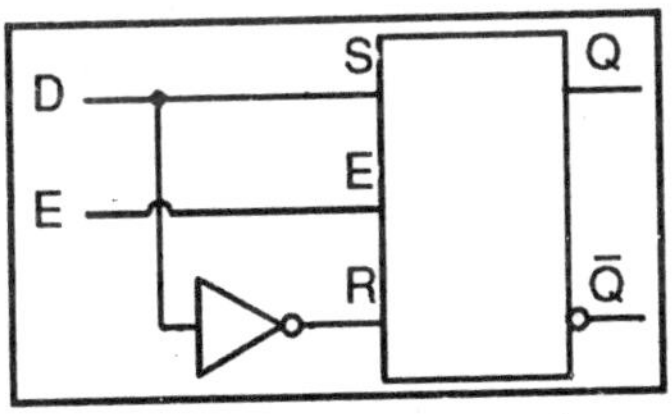

Like both the S-R and gated S-R latches, the D latch circuit may be found as its own prepackaged circuit, complete with a standard symbol:

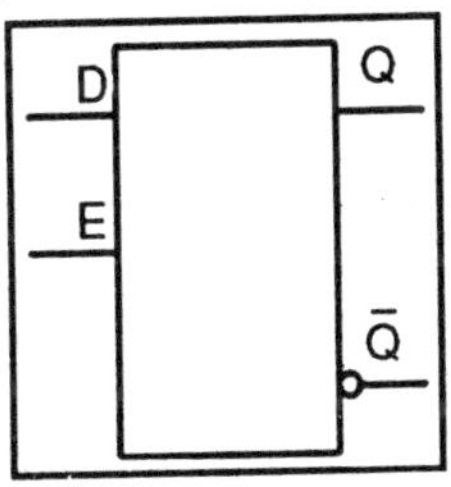

The D latch is nothing more than a gated S-R latch with an inverter added to make R the complement (inverse) of S. Let's explore the ladder logic equivalent of a D latch, modified from the basic ladder diagram of an S-R latch:

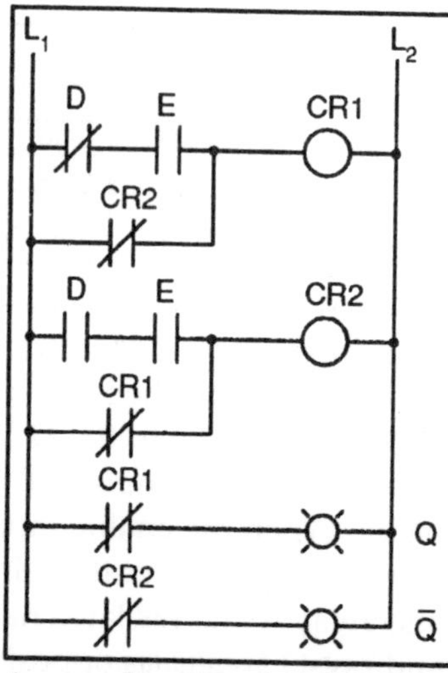

E	D	Q	$\bar{Q}$
0	0	Latch	Latch
0	1	Latch	Latch
1	0	0	1
1	1	1	0

An application for the D latch is a 1-bit memory circuit. You can "write" (store) a 0 or 1 bit in this latch circuit by making the enable input high (1) and setting D to whatever you want the stored bit to be. When the enable input is made low (0), the latch ignores the status of the D input and merrily holds the stored bit value, outputting at the stored value at Q, and its inverse on output not-Q.

ASYNCHRONOUS CIRCUIT

An asynchronous sequential circuit is a sequential circuit whose behaviour depends only on the order in which its input signals change and can be affected at any instant of time. In addition to the FFs, a register may have combinational gates that control when and how new information is transferred into the register.

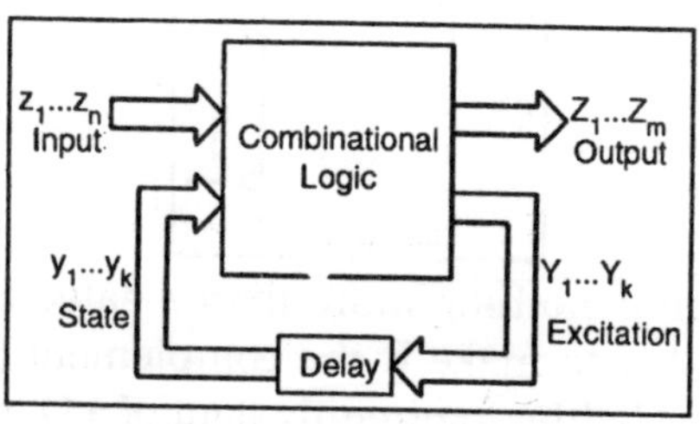

Fig. Asynchronous Sequential Circuit Model.

- $X = (x_1, x_2\ x_3, ..., x_n)$: n input Variables
- $z = (z_1, z_2, ..., z_m)$: m Output Variables
- $y = (y_1, y_2, ..., y_k)$: k State Variables (Present State)
- $Y = (Y_1, Y_2, ..., Y_k)$: k Excitation Variables

In the above circuit, the state transition occurs when there is an input change (no clock pulses). Memory (delay) elements are either latches (unclocked) or time-delay elements (instead of clocked FFs as in a synchronous sequential circuit). An asynchronous sequential circuit quite often resembles a combinational circuit with feedback. Faster and often cheaper than synchronous ones, but more difficult to design, verify, or test (due to possible timing problems involved in the feedback path).

- *Steady-state condition*: To ensure proper operation, simultaneous changes of 2 or more input variables are usually prohibited.
- *Fundamental-mode operation*: Only one input variable can change at any time, and the time between 2 input changes must be longer than the time it takes the circuit to reach a stable state.

ANALYSIS

The analysis consists of obtaining a table or a diagram that describes the sequence of internal states and outputs as a function of changes in the input variables.

- Determine all feedback loops.
- Designate each feedback-loop output with Yi and its corresponding input with yi for i = 1;:::; k, where k is the number of feedback loops.
- Derive the boolean functions for all Y 's.
- Plot the transition table from the equations.

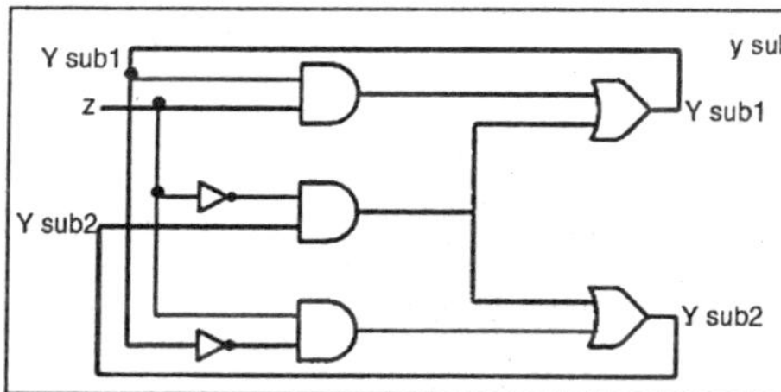

y sub 1 ysub2	0	1
00	(00)	01
01	11	(01)
11	(11)	10
10	00	(10)

Transition Table

Excitation Variables: Y_1 and Y_2

Secondary Variables: y_1 and y_2

$$Y_1 = xy_1 + x'y_2$$

$$Y_1 = xy_1' + x'y_2$$

The delay associated with each feedback loop is obtained from the propagation delay between each y input and its corresponding Y output.

In an asynchronous sequential circuit, the internal state can change immediately after a change in the input.

- *Total state*: Internal state + input value.
- *Flow table*: A transition table in which states are named by letter symbols instead of specific binary values.

Stable State

- *Primitive flow table*: one that has only one stable state in each row. To obtain the circuit described by a flow table, it is necessary to assign to each state a distinct binary value, which converts the flow table into a transition table.
- *Race condition*: when 2 or more binary state variables change value in response to a change in an input variable.

  May cause the state variables to change in an unpredictable manner.
- *Noncritical race*: final stable state does not depend on the order in which the state variables change.
- *Critical race*: final stable state depends on the order in which the state variables change.

For proper operation, critical races must be avoided.

Races may be avoided by making a proper state assignment. Races can be avoided by directing the circuit through intermediate unstable states with a unique state-variable change.

FEATURES

- Obtain a primitive flow table.
- Reduce the flow table.

- Assign binary state variables to obtain the transition table.
- Assign output values to the dashes to obtain the output maps.
- Simplify the excitation and output functions.
- Draw the logic diagram.

We will design a gated latch circuit with 2 inputs, G (gate) and D (data), and one output, Q. The binary value at the D input is transferred to the Q output when and only when G = 1. When G falls to 0, the latch retains the value at the Q output, which does not change even if D changes.

The table of total states and the corresponding primitive flow table is shown below.

State	Inputs D	Inputs G	Output Q	Comments
a	0	1	0	D = Q
b	1	1	1	D = Q
c	0	0	0	After a or d
d	1	0	0	After c
e	1	0	1	After b or f
f	0	0	1	After e

Table of Total States

	DG 00	01	11	10
a	c,–	(a)0	b,–	–,–
b	–,–	a,–	(b)1	e,–
c	(c)0	a,–	–,–	d,–
d	c,–	–,–	b,–	(d)0
e	f,–	–,–	b,–	(e)1
f	(f)1	a,–	–,–	e,–

Primitive Flow Table — States to be Merged

The next step is to reduce the primitive flow table, which is shown below. Then,

	DG 00	01	11	10
a,b,c	(c)0	(a)0	b,–	(a)0
b,e,f	(f)1	a,–	(b) 1	(e) 1

→

	DG 00	01	11	10
a	(a) 0	(a) 0	b,–	(a) 0
b	(b) 0	a,–	(b) 0	(b)0

following state assignment, we convert the flow table into a transition table (by assigning a = 0 and b = 1). From the table, the simplified O/P function (Q) and the excitation function (Y) with respect to D, G, and y can be derived using the K-map method. Note that the don't-cares are assigned such that

$$Q = Y$$

$Y = DG + G' y;$ $\qquad\qquad Q = Y$

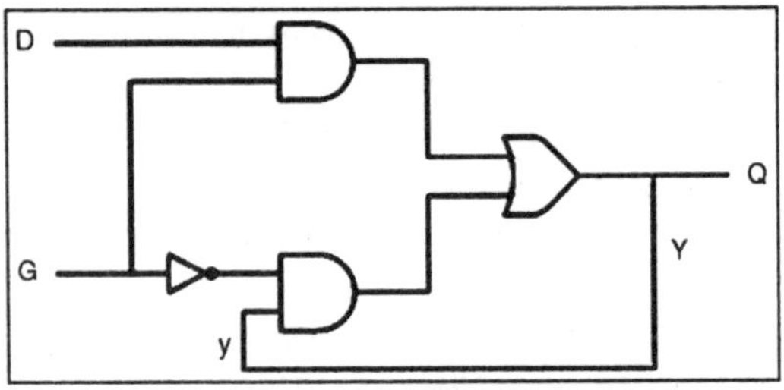

REDUCTION OF STATE

The procedure for reducing the number of internal states in an asynchronous sequential circuit resembles the procedure that is used for synchronous circuits.

IMPLICATION TABLE

The state-reduction procedure for completely specified state tables is based on the algorithm that two states in a state table can be combined into one if they can be shown to be equivalent. There are occasions when a pair of states do not have the same next states, but, nonetheless, go to equivalent next states.

Consider the following state table:

Present State	Next State		output	
	x = 0	x = 1	x = 0	x = 1
a	c	b	0	1
b	d	a	0	1
c	a	d	1	0
d	b	d	1	0

(a, b) imply (c, d) and (c, d) imply (a, b). Both pairs of states are equivalent; i.e., a and b are equivalent as well as c and d. The checking of each pair of states for possible equivalence in a table with a large number of

states can be done systematically by means of an implication table. This a chart that consists of squares, one for every possible pair of states, that provide spaces for listing any possible implied states.

Consider the following state table:

Present State	Next State		output	
	x = 0	x = 1	x = 0	x = 1
a	d	6	0	0
b	e	a	0	0
c	g	f	0	1
d	a	d	1	0
e	a	d	1	0
f	c	b	0	0
g	a	e	1	0

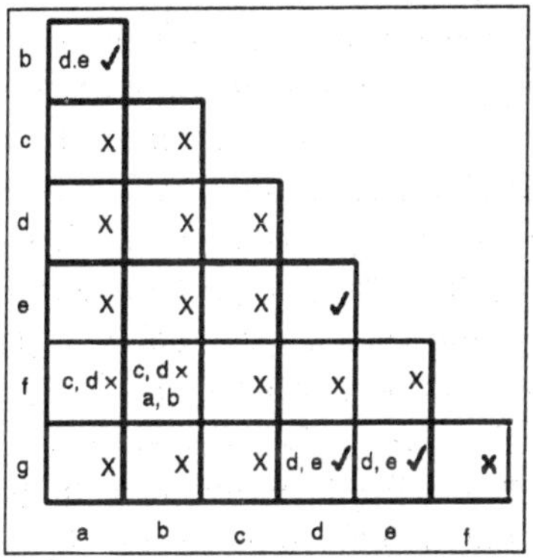

On the left side along the vertical are listed all the states defined in the state table except the last, and across the bottom horizontally are listed all the states except the last. The states that are not equivalent are marked with a 'x' in the corresponding square, whereas their equivalence is recorded with a '?'. Some of the squares have entries of implied states that must be further investigated to determine whether they are equivalent or not.

The step-by-step procedure of filling in the squares is as follows:

- Place a cross in any square corresponding to a pair of states whose outputs are not equal for every input.
- Enter in the remaining squares the pairs of states that are implied by the pair of states representing the squares. We do that by starting from the top square in the left column and going down and then proceeding with the next column to the right.
- Make successive passes through the table to

determine whether any additional squares should be marked with a 'x'. A square in the table is crossed out if it contains at least one implied pair that is not equivalent.

- Finally, all the squares that have no crosses are recorded with check marks. The equivalent states are: (a, b), (d, e), (d, g), (e, g).

We now combine pairs of states into larger groups of equivalent states. The last three pairs can be combined into a set of three equivalent states (d, e, g) because each one of the states in the group is equivalent to the other two. The final partition of these states consists of the equivalent states found from the implication table, together with all the remaining states in the state table that are not equivalent to any other state:

(a, b) (c) (d, e, g) (f)

The reduced state table is:

Present State	**Next State**		**output**	
	x = 0	**x = 1**	**x = 0**	**x = 1**
a	d	a	0	0
c	d	f	0	1
d	a	d	1	0
f	a	c	0	0

FLOW TABLE

There are occasions when the state table for a sequential circuit is incompletely specified.

Incompletely specified states can be combined to reduce the number of states in the flow table. Such states cannot be called equivalent, but, instead they are said to be compatible.

The process that must be applied in order to find a suitable group of compatibles for the purpose of merging a flow table is divided into three steps:

- Determine all compatible pairs by using the implication table.
- Find the maximal compatibles using a merger diagram.

- Find a minimal collection of compatibles that covers all the states and is closed.

We will now proceed to show and explain the three procedural steps using the following primitive flow table:

	DG 00	01	11	10
a	C, –	(a) 0	b, –	–, –
b	–, –	a, –	(b) 1	e, –
c	(c) 0	a, –	–, –	d, –
d	c, –	–, –	b, –	(d) 0
e	f, –	–, –	b, –	(e) 0
f	(f) 0	a, –	–, –	e, –

COMPATIBLE PAIRS

Two states are compatible if in every column of the corresponding rows in the flow table, they are identical or compatible states and if there is no conflict in the output values.

The compatible pairs (?) are:

(a, b) (a, c) (a, d) (b, e) (b, f) (c, d) (e, f)

	a	b	c	d	e
b	✓				
c	✓	d, e ×			
d	✓	d, e ×	✓		
e	C, f ×	✓	d, e × c, f ×	×	
f	C, f ×	✓	×	d, e × C, f ×	✓

MAXIMAL COMPATIBLES

The maximal compatible is a group of compatibles that contains all the possible combinations of compatible states. The maximal compatible can be obtained from a merger diagram:

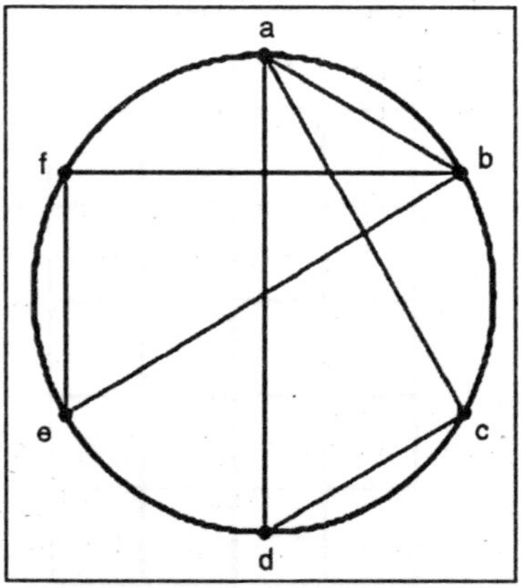

The above merger diagram is obtained from the list of compatible pairs derived from the previous implication table. A line represents a compatible pair. A triangle constitutes a compatible with three states.

The maximal compatibles are:

(a, b) (a, c, d) (b, e, f)

In the case where a state is not compatible to any other state, an isolated dot represents this state.

DIGITAL SYSTEM

While the binary numeration system is an interesting mathematical abstraction, we haven't yet seen its practical application to electronics. What makes binary numeration so important to the application of digital electronics is the ease in which bits may be represented in physical terms. Because a binary bit can only have one of two different values, either 0 or 1, any physical medium capable of switching between two saturated states may be used to represent a bit. Consequently, any physical system capable of representing binary bits is able to represent numerical quantities, and potentially has the ability to manipulate those numbers.

Electronic circuits are physical systems that lend themselves well to the representation of binary numbers. Transistors, when operated at their bias limits, may be in one of two different states: either cutoff.

If a transistor circuit is designed to maximize the probability of falling into either one of these states (and not operating in the linear, or active, mode), it can serve as a physical representation of a binary bit.

A voltage signal measured at the output of such a circuit may also serve as a representation of a single bit, a low voltage representing a binary "0" and a (relatively) high voltage representing a binary "1." Consider the following transistor circuit:

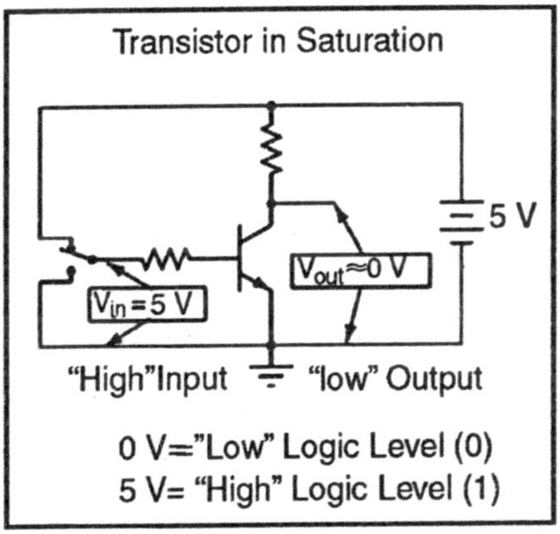

In this circuit, the transistor is in a state of saturation by virtue of the applied input voltage (5 volts) through the two-position switch. Because it's saturated, the transistor drops very little voltage between collector and emitter, resulting in an output voltage of (practically) 0 volts.

If we were using this circuit to represent binary bits, we would say that the input signal is a binary "1" and that the output signal is a binary "0." Any voltage close to full supply voltage (measured in reference to ground, of course) is considered a "1" and a lack of voltage is considered a "0." Alternative terms for these voltage levels are high (same as a binary "1") and low (same as a binary "0").

A general term for the representation of a binary bit by a

circuit voltage is logic level. Moving the switch to the other position, we apply a binary "0" to the input and receive a binary "1" at the output:

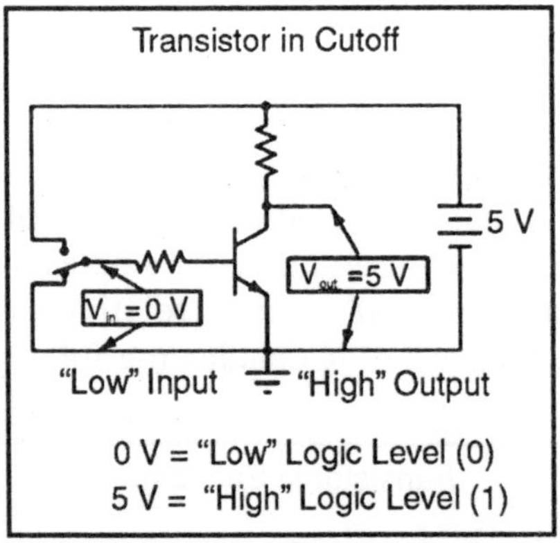

What we've created here with a single transistor is a circuit generally known as a logic gate, or simply gate. A gate is a special type of amplifier circuit designed to accept and generate voltage signals corresponding to binary 1's and 0's. As such, gates are not intended to be used for amplifying analog signals (voltage signals between 0 and full voltage).

Used together, multiple gates may be applied to the task of binary number storage (memory circuits) or manipulation (computing circuits), each gate's output representing one bit of a multi-bit binary number.

The gate shown here with the single transistor is known as an inverter, or NOT gate, because it outputs the exact opposite digital signal as what is input. For convenience, gate circuits are generally represented by their own symbols rather than by their constituent transistors and resistors. The following is the symbol for an inverter:

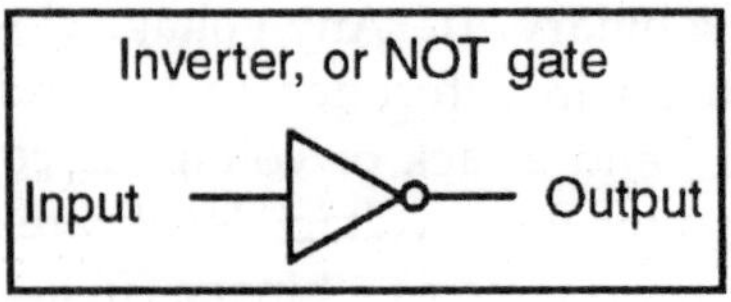

An alternative symbol for an inverter is shown here:

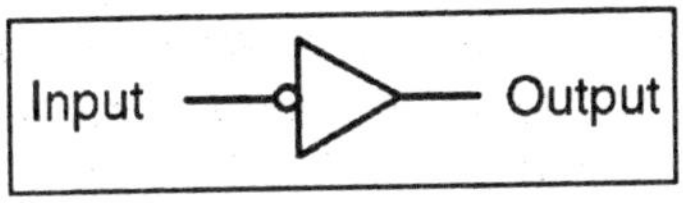

Notice the triangular shape of the gate symbol, much like that of an operational amplifier. As was stated before, gate circuits actually are amplifiers. The small circle, or "bubble" shown on either the input or output terminal is standard for representing the inversion function.

As you might suspect, if we were to remove the bubble from the gate symbol, leaving only a triangle, the resulting symbol would no longer indicate inversion, but merely direct amplification. Such a symbol and such a gate actually do exist, and it is called a buffer, the subject of the next section.

Like an operational amplifier symbol, input and output connections are shown as single wires, the implied reference point for each voltage signal being "ground." In digital gate circuits, ground is almost always the negative connection of a single voltage source (power supply).

Dual, or "split," power supplies are seldom used in gate circuitry. Because gate circuits are amplifiers, they require a source of power to operate. Like operational amplifiers, the power supply connections for digital gates are often omitted from the symbol for simplicity's sake. If we were to show all the necessary connections needed for operating this gate, the schematic would look something like this:

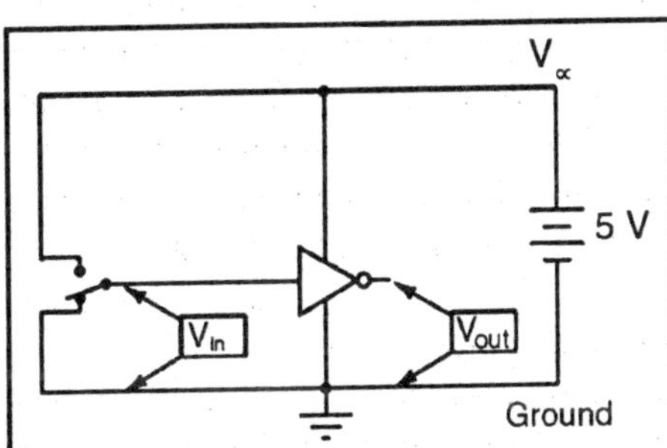

Power supply conductors are rarely shown in gate circuit schematics, even if the power supply connections at each gate are. Minimizing lines in our schematic, we get this:

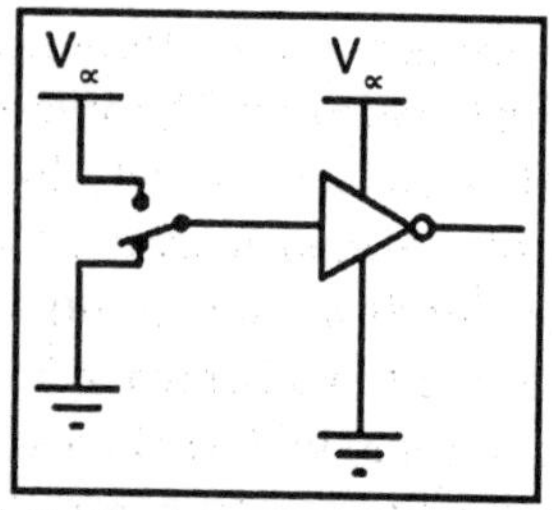

"V_{cc}" stands for the constant voltage supplied to the collector of a bipolar junction transistor circuit, in reference to ground. Those points in a gate circuit marked by the label "V_{cc}" are all connected to the same point, and that point is the positive terminal of a DC voltage source, usually 5 volts.

As we will see in other sections of this chapter, there are quite a few different types of logic gates, most of which have multiple input terminals for accepting more than one signal. The output of any gate is dependent on the state of its input(s) and its logical function.

One common way to express the particular function of a gate circuit is called a truth table. Truth tables show all combinations of input conditions in terms of logic level states (either "high" or "low," "1" or "0," for each input terminal of the gate), along with the corresponding output logic level, either "high" or "low." For the inverter, or NOT, circuit just illustrated, the truth table is very simple indeed:

NOT Gate Truth Table

Input  Output

Input	Output
0	1
1	0

Truth tables for more complex gates are, of course, larger than the one shown for the NOT gate. A gate's truth table must have as many rows as there are possibilities for unique input combinations. For a single-input gate like the NOT gate, there are only two possibilities, 0 and 1. For a two input gate, there are four possibilities (00, 01, 10, and 11), and thus four rows to the corresponding truth table. For a three-input gate, there are eight possibilities (000, 001, 010, 011, 100, 101, 110, and 111), and thus a truth table with eight rows are needed. The mathematically inclined will realise that the number of truth table rows needed for a gate is equal to 2 raised to the power of the number of input terminals.

HIGH VACUUM DIODE

A vacuum diode is an electron valve that supports current flow through it in only one direction. It is composed of two electrical "parts" or elements and the evacuated glass envelope. Additionally, there is usually a heater, a coil of special alloy wire that can be driven to red heat by a modest application of voltage. (A vacuum diode without a heater would be a cold cathode diode.) The heater is coiled inside a cylindrical element called the cathode. The cathode is composed of a special alloy, or, more properly is coated with an alloy, that has high thermal emission characteristics. That means that if it's heated sufficiently, electrons actually leave the metallic crystal structure of the metal (boil off) and hang around outside in the space nearby forming what is called a space charge. When we built the tube, we pumped all the air out, sealed the glass, and then fired a "getter" inside the tube to chemically bind the remaining bit of gas that was there. The tube really is highly evacuated. That's important because we don't want all those light weight electrons bumping into gas atoms and molecules when they are hanging around in the space charge, and especially when transiting the gap between the cathode and the other element inside, the anode or plate. If a voltage is applied across the cathode and plate, and the plate is negative, no current will flow in the tube. Electrons flow from negative to positive, but, darn it, they can't leave the plate to go to the cathode. That's called reverse biased. It's hooked up "backwards" like that. But if the cathode is connected to negative and the positive connected to the plate, electrons that were in the space charge will be driven away from the cathode and attracted to the plate. Current will flow in this direction. That's forward bias.

The hot cathode emits electrons in droves. Many more than are needed. They build up in the volume around the cathode and form a large pool of electrons just waiting for something positive to happen. This pool of electrons is called the space charge. When the plate (anode) is made positive some of the electrons are attracted to it. They impact on it, are absorbed into the metal and electrons flow out of the plate connection into the battery or what ever provided the positive voltage. The other end of the voltage source must be connected back to the cathode in some way.

If the polarity is reversed which makes the plate negative with respect to the cathode the electrons in the space charge are repelled away from the plate and no conduction takes place. The plate is cold and is made of a metal that is a very poor emitter of electrons. The voltages are not nearly as high as those used by Crooks so there is no cold emission from the plate. This makes the diode conduct current in only one direction.

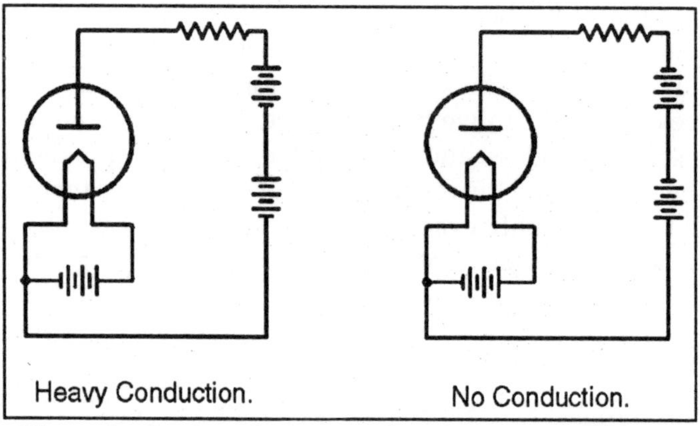

The cathode of a tube is coated with a very special material that emits electrons very easily. It can readily be damaged by operating above or below the proper temperature or by being "poisoned" by the residual air in the tube. The space charge serves another vital function. There is no such thing as a perfect vacuum. In fact the space in near earth orbit is a better vacuum than can be made on the surface of earth. Farther out in space the vacuum is even better. What this means is there are lots of air molecules running around inside of a vacuum tube. The space charge protects the cathode from these air molecules. Some of these molecules lose an electron and become positive ions. They are attracted to the negative cathode and if it weren't for the space charge they would impact on the surface of the cathode doing considerable damage. Over time the ability of the cathode to emit electrons would be seriously impaired rendering the tube useless. As these ions run through the space charge they gain electrons from the many collisions and become negative ions. Their direction is reversed turning them away from the delicate cathode.

INDIRECTLY HEATED CATHODES

Most vacuum tubes have a basically cylindrical configuration. Figure two shows how a filament (directly heated) cathode type diode is made along with a heater type (indirectly heated) cathode type diode.

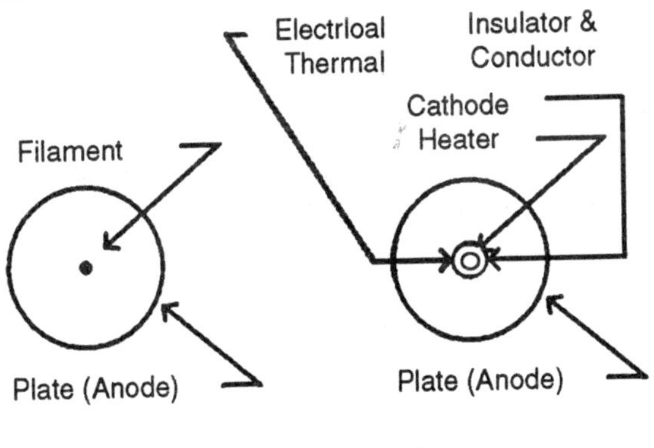

End View of Cylindrical Constructs.

PHOTODIODES

DEFINITION

Photodiodes are semiconductor light sensors that generate a current or voltage when the P-N junction in the semiconductor is illuminated by light. The term photodiode can be broadly defined to include even solar batteries, but it usually refers to sensors used to detect the intensity of light. Photodiodes can be classified by function and construction as follows:

VARIETY

- PN photodiode
- PIN photodiode
- Schottky type photodiode
- APD (Avalanche photodiode)

FEATURES

- Excellent linearity with respect to incident light
- Low noise
- Wide spectral response
- Mechanically rugged
- Compact and lightweight
- Long life

OPERATIONS

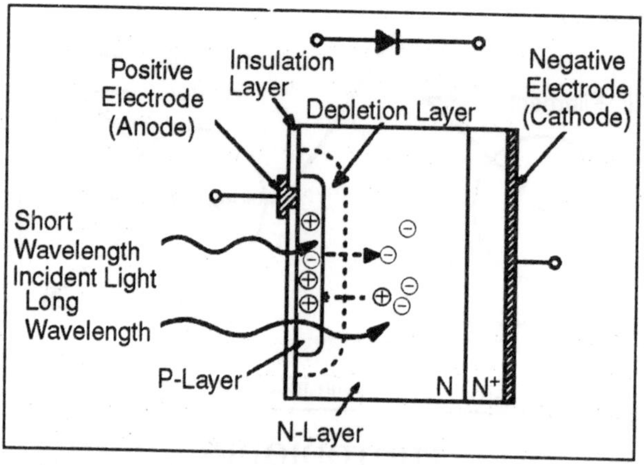

Fig. Below Shows a Cross Section of a Photodiode.

The P-layer material at the active surface and the N material at the substrate form a PN junction which operates as a photoelectric converter. The usual P-layer for a Si photodiode is formed by selective diffusion of boron, to a thickness of approximately 1 μm or less and the neutral region at the junction between the P- and N-layers is known as the depletion layer. By controlling the thickness of the outer P-layer, substrate N-layer and bottom N^+-layer as well as the doping concentration, the spectral response and frequency response can be controlled. When light strikes a photodiode, the electron within the crystal structure becomes stimulated. If the light energy is greater than the band gap energy Eg, the electrons are pulled up into the conduction band, leaving holes in their place in the valence band as shown below:

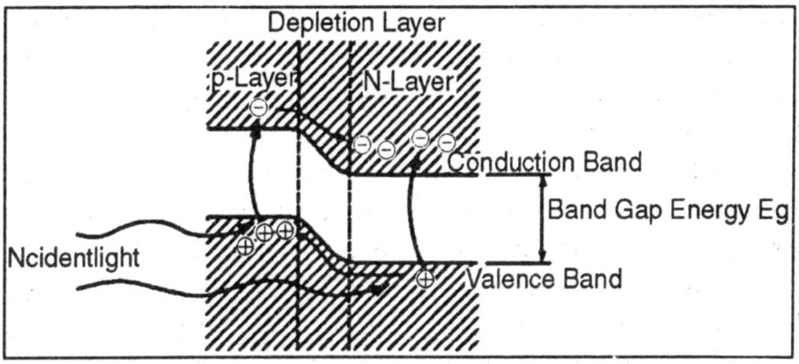

These electron-hole pairs occur throughout the P-layer, depletion layer and N-layer materials. In the depletion layer the electric field accelerates these electrons toward the N-layer and the holes toward the P-layer. Of the electron-hole pairs generated in the N-layer, the electrons, along with electrons that have arrived from the P-layer, are left in the N-layer conduction band. The holes at this time are being diffused through the N-layer up to the depletion layer while being accelerated, and collected in the P-layer valence band. In this manner, electron-hole pairs which are generated in proportion to the amount of incident light are collected in the N- and P-layers.

Index